Seasonality has effects on a wide range of human functions and activities, and is important in the understanding of human–environment relationships. In this volume, distinguished contributors, including human biologists, anthropologists, physiologists and nutritionists, consider many of the different ways in which seasonality influences human biology and behaviour.

Topics addressed include the influence of seasonality on hominid evolution, seasonal climatic effects on human physiology, fertility and physical growth, seasonality in morbidity, mortality and nutritional state, and seasonal factors in food production, modernisation and work organisation in Third World economies.

This book will be of interest to graduate students and researchers in human biology, anthropology and nutrition.

*SOCIETY FOR THE STUDY OF HUMAN BIOLOGY*
*SYMPOSIUM SERIES: 35*

Seasonality and human ecology

# PUBLISHED SYMPOSIA OF THE
# SOCIETY FOR THE STUDY OF HUMAN BIOLOGY

*Numbers 1–9 were published by Pergamon Press, Headington Hill Hall, Headington, Oxford OX3 0BY. Numbers 10–24 were published by Taylor & Francis Ltd, 10–14 Macklin Street, London WC2B 5NF. Further details and prices of back-list numbers are available from the Secretary of the Society for the Study of Human Biology.*

# Seasonality and human ecology

*35th Symposium Volume of the*
*Society for the Study of Human Biology*

EDITED BY

## S.J. ULIJASZEK

*Department of Biological Anthropology*
*University of Cambridge*

AND

## S.S. STRICKLAND

*Centre for Human Nutrition*
*London School of Hygiene and Tropical Medicine*
and
*Department of Anthropology*
*University College London*

 CAMBRIDGE
UNIVERSITY PRESS

Published by the Press Syndicate of the University of Cambridge
The Pitt Building, Trumpington Street, Cambridge CB2 1RP
40 West 20th Street, New York, NY 10011–4211, USA
10 Stamford Road, Oakleigh, Melbourne 3166, Australia

© Cambridge University Press 1993

First published 1993

Printed in Great Britain at the University Press, Cambridge

*A catalogue record for this book is available from the British Library*

*Library of Congress cataloguing in publication data*

Seasonality and human ecology: 35th symposium volume of the Society
for the Study of Human Biology / edited by S.J. Ulijaszek and S.S.
Strickland.
    p. cm. – (Society for the Study of Human Biology Symposium
series ; 35)
  Includes index.
  1. Man – Influence of nature – Congresses. 2. Chronobiology –
Congresses. 3. Human biology – Congresses. 4. Seasons – Congresses.
I. Ulijaszek, Stanley J. II. Strickland S.S. III. Series.
GF51.S46 1993
304.2–dc20    92–43449  CIP

ISBN 0 521 43147 6 hardback

# Contents

# Contributors

E. Bénéfice
Nutrition Unit, Institut Francais de Recherche Scientifique pour le
Developpement en Coopération (ORSTOM), Centre ORSTOM de
Dakar, BP 1386, Dakar, Senegal

A.H. Bittles
Department of Anatomy and Human Biology, King's College
London, Strand, London WC2R 2LS, UK, and Edith Cowan
University, Perth, Western Australia

F. Branca
Istituto Nazionale della Nutrizione, Via Ardeatina 546 00178 Rome,
Italy

A. Chahnazarian
Population and Health Unit, Institut Francais de Recherche
Scientifique pour la Developpement en Coopération (ORSTOM),
Centre ORSTOM de Dakar, BP 1386, Dakar, Senegal

T.J. Cole
Dunn Nutritional Laboratory, Downhams Lane, Milton Road,
Cambridge CB4 1XJ, UK

K.J. Collins
Division of Geriatrics, Department of Medicine, University College
and Middlesex School of Medicine, St Pancras Hospital, 4 St Pancras
Way, London NW1 0PE, UK

M.J. Dauncey
Department of Molecular and Cellular Physiology, AFRC Institute of
Animal Physiology and Genetics Research, Babraham, Cambridge
CB2 4AT, UK

V. Delaunay
Population and Health Unit, Institut Francais de Recherche
Scientifique pour le Developpement en Coopération (ORSTOM),
Centre ORSTOM de Dakar, BP 1386, Dakar, Senegal

A. Ferro-Luzzi
Istituto Nazionale della Nutrizione, Via Ardeatina 546 00178 Rome,

Italy

R.A. Foley
Department of Biological Anthropology, University of Cambridge,
Downing Street, Cambridge CB2 3DZ, UK

I. de Garine
Centre National de la Recherche Scientifique, 31 Bis, Rue de Sevres,
92100 Boulogne, Paris, France

R.A. Huss-Ashmore
Department of Anthropology, University of Pennsylvania, University
Museum, Philadelphia PA 19104–6398, USA

D.L. Ingram
Department of Molecular and Cellular Physiology, AFRC Institute of
Animal Physiology and Genetics Research, Babraham, Cambridge
CB2 4AT, UK

F.E. Johnston
Department of Anthropology, University of Pennsylvania, University
Museum, Philadelphia PA 19104–6398, USA

C. Panter-Brick
Department of Anthropology, University of Durham, 43 Old Elvet,
Durham DH1 3NH, UK

L. Rosetta
Laboratoire de Physiologie des Adaptations, Université Paris 5, 24
Rue du Faubourg Saint-Jacques, 75015 Paris, France

L. Sanz
Department of Anthropology, Complutense University, Madrid,
Spain

F. Simondon
Nutrition Unit, Institut Francais de Recherche Scientifique pour le
Developpement en Coopération (ORSTOM), Centre ORSTOM de
Dakar, BP 1386, Dakar, Senegal

K.B. Simondon
Nutrition Unit, Institut Francais de Recherche Scientifique pour le
Developpement en Coopération (ORSTOM), Centre ORSTOM de
Dakar, BP 1386, Dakar, Senegal

S.S. Strickland
Centre for Human Nutrition, London School of Hygiene and
Tropical Medicine, Keppel Street, London WC1E 7HT, and
Department of Anthropology, University College London, Gower

Street, London WC1, UK

M.A. Stroud
Applied Physiology (A), Army Personnel Research Establishment,
Ministry of Defence, Farnborough, Hants. GU14 6TD, UK

A.M. Tomkins
Centre for International Child Health, Institute of Child Health, 30
Guilford Street, London WC1N 1EH, UK

S.J. Ulijaszek
Department of Biological Anthropology, University of Cambridge,
Downing Street, Cambridge CB2 3DZ, UK

# Acknowledgements

We thank Professor G.A. Harrison and Professor F.E. Johnston, who shared with the Editors the chairing of the Symposium sessions, and the Royal Society, British Academy and British Council for financial support.

# 1  *Introduction*

S.J. ULIJASZEK AND S.S. STRICKLAND

Human ecology is the study of the interrelations that exist between individuals, populations, and the ecosystems of which they are a part, while seasonality has been defined as regular recurring intra-annual fluctuation of environments, and of the individuals and populations living in them. The appreciation of seasonality as a significant source of human variability has a considerable history, anthropologists in particular long having recognised the importance of season in the structuring of people's lives and of their work and ritual activities. However, it is only more recently that the effects of seasonality on a range of human functions and activities have been widely investigated.

The Society for the Study of Human Biology Symposium on Seasonality and Human Ecology, held on April 9–10, 1992 at the University of Cambridge, allowed the different ways in which seasonality influences human biology and behaviour to be examined in a systematic manner by human biologists, anthropologists, physiologists and nutritionists. The papers discussed at that meeting are published in this volume.

In the first contribution (Chapter 2), F.E. Johnston discusses the range of biological phenomena associated with seasonality and concludes that, as a species, humans are enormously responsive to the seasonal changes and environmental cycles that characterise their ecosystems. He argues that environmental seasonality is not so much a stress to which contemporary human populations must adapt, but rather a basic component of the ecosystems within which our ancestors evolved, one that has left its mark indelibly upon the makeup of our biology and behaviour. By using three examples of seasonal effects upon human biology: 1. growth and development; 2. conception and birth; and 3. food intake and nutrition, he is able to make three generalisations. The first is that a wide range of human biological processes show periodicities that seem to exist apart from exogenous influences; the second is that periodic processes in humans are biobehavioural in nature; and the third is that human groups are enormously sensitive to their ecosystems. These generalisations are examined and developed in subsequent chapters.

R.A. Foley (Chapter 3) develops the theme that seasonality is a major

1

evolutionary pressure, using variation in East African seasonality as a basis for considering the diverse adaptive trends of early African hominids. This analysis is extended by examining some of the problems faced by tropical species adapting to temperate seasonality, including thermoregulation. He concludes that one of the principal proximate ecological factors influencing the success of hominids in their radiation is likely to have been the time available for foraging during mid- to high-latitude winters.

The importance of thermoregulation in adaptation to seasonal environments is made clear by M.A. Stroud (Chapter 4), who reviews current knowledge about physiological responses to hot and cold environments in modern humans. He highlights areas of contention in our understanding of human responses to cold stress, including the importance of shivering, and of non-shivering thermogenesis. On the basis of recent evidence, he raises the possibility that non-shivering thermogenesis may be an important adaptation in adults to severe cold.

Another way in which seasonal environments may operate directly on human physiology is through variation in daylength. In humans, there is every reason to believe that light signals received by the eye have effects on the pineal gland similar to those in animals showing obvious seasonal rhythms. Reviewing this topic, D.L. Ingram and M.J. Dauncey (Chapter 5) conclude that the intensity of light needed to reduce the secretion of melatonin or reset a circadian rhythm is much greater in humans than in other animals, and that such signals have little if any effect on the control of human reproductive function. They show that the seasonality of birth-rates and conception reported for populations in many parts of the world have nothing to do with hormonal changes associated with changing daylength.

L. Rosetta (Chapter 6) continues the theme of seasonality and fertility by examining the effects of photoperiod, temperature and humidity on human sperm quality, hormonal rhythm, menstrual pattern and birth seasonality. Although many of the mechanisms and mediators involved in the seasonal regulation of fertility are not known, seasonal variation in sperm quality and the duration of the menstrual cycle are some of the most important factors influencing birth-rate in all populations. However, in Third World communities seasonality of rainfall and temperature are additional factors influencing fertility, through the physiological effects of changing energy expenditure and nutritional status on the release of hormones involved in the control of ovulation in women.

Seasonality in fertility is placed in an ecological context by S.J. Ulijaszek (Chapter 7), who examines the relationships between birth-rate, pregnancy outcome, lactational performance and child mortality in a rural community in The Gambia, where seasonality of food intake, energy expenditure and infectious disease are experienced. He shows that seasonal troughs in birth-rate are more closely related to seasonally low intakes of dietary energy, and that childhood mortality is lowest in children conceived at the time of year when fertility is lowest.

From reproductive performance, the focus shifts to seasonal effects on physical growth and development. T. J. Cole (Chapter 8) carries out an analysis of infant growth data from two contrasting populations, one in rural Gambia, the other in Cambridge, England. There is seasonality of growth in both groups, the effect being greater in the Gambian than the Cambridge infants. In The Gambia, birth during certain times of the year is better for subsequent growth and development than birth at other times.

In the next chapter, A.H. Bittles and L. Sanz (Chapter 9) consider the possibility that polygenic multifactorial disorders show seasonal variation. Analysis of British data on neural tube defects and schizophrenia at birth reveals seasonality in both. They suggest that seasonality in children born with neutral tube defects is environmentally mediated through the folate nutritional status of their mothers during pregnancy.

A. Tomkins (Chapter 10) reviews the evidence for the influence of a range of factors on seasonality of a number of infectious diseases, including dysentery, malaria, measles, respiratory infections and diarrhoea. He highlights the complexity of disease ecology in the tropics, upon which seasonality is superimposed.

By contrast, K.J. Collins (Chapter 11) examines seasonal morbidity and mortality in temperate regions, focussing on the elderly in Britain. He shows that the characteristic cold weather pattern of increased winter morbidity and mortality in cold and temperate regions is more pronounced in Britain than in many comparable countries, hypothermia contributing only a tiny proportion of excess winter deaths. Respiratory mortality in winter is most closely associated with influenza epidemics and levels of home heating, while outdoor cold is an important factor in the aetiology of winter cardiac deaths.

The next chapter is the first of several to consider seasonal aspects of food and nutrition. A. Ferro-Luzzi and F. Branca (Chapter 12) attempt to determine the proportion of the world's population exposed to seasonal undernutrition. They use an index of seasonality in association with body mass index to identify the parts of the world that are at high risk of seasonal nutritional stress. They conclude that: 1. populations in parts of sub-Sahelian and Southern Africa and in India are at greatest risk; 2. populations in China are at considerably lower risk; and 3. Latin American populations are at very low risk. Although the model developed by them is somewhat speculative, and despite the acknowledgement that its utility is restricted by inadequate data for many parts of the world, it is a method that may be of considerable use to nutritionists and planners in the future.

In contrast to the global approach in the previous chapter, K.B. Simondon, E. Bénéfice, F. Simondon, V. Delaunay and A. Chahnazarian (Chapter 13) illustrate the effects of climatic seasonality on the nutritional status of adults and children in two populations in Senegal, West Africa, one having access to modern irrigation technology, the other not. They show how modern irrigation agriculture has served to reduce seasonality in nutritional status in women, but not in children.

I. de Garine (Chapter 14) considers the perception of seasonality in two other African communities, the Massa and Mussey, and how this relates to climate and the human biological effects of seasonality. He shows how complex the biobehavioural relations between humans and their seasonal environment can be. It is important to note that seasonally high biological stress does not necessarily coincide with seasonally high psychological stress in these societies.

R.A. Huss-Ashmore (Chapter 15) addresses the impact of economic change in developing countries on the ability of traditional farmers to cope with seasonality, concluding that agricultural development may increase incomes and food consumption, but does not do away with their seasonality. This chapter illustrates the fact that agricultural development is one of an array of strategies for procuring food and other resources, its adaptive success estimated by its ability to provide adequate resources during the lean season while not jeopardising health or long-term productivity.

In the final chapter (Chapter 16), C. Panter-Brick examines the ways in which communities living in seasonal environments organise themselves to accomplish their subsistence tasks, and the underlying logic governing their choice of work patterns. She shows that in Nepal and elsewhere, households combine to work in labour groups at times of peak work-load, thus enabling them to complete urgent tasks and to maximise labour efficiency for tasks that require cooperation and pooled resources. At other times, cooperative strategies may be less important, and so are less likely to be practised.

# 2 *Seasonality and human biology*

FRANCIS E. JOHNSTON

Human biology is the study of the dynamics of biological variation in human populations. Of the numerous research avenues followed by human biologists one of the most challenging is the description and analysis of the relationships between the environmental milieux and the biobehavioural structures of human groups. The challenge lies not only in the specification and measurement of biological and environmental variables, but also the analysis of their interactions with each other within the contexts of hypothesis testing and the construction of models. Neither the biological states of populations and organisms, nor the surrounding environment, are constant, and it is this dual lack of constancy that creates problems of methodology and interpretation in research.

Seasonality, defined by Huss-Ashmore (1988a, p. 5) as 'regular, recurring intra-annual fluctuation' is of particular significance to human biologists when analyzing the effects of the environment. The individual, the population, and the environment all may fluctuate. And their fluctuations may be stable, i.e., predictable (see Bloom, 1964), or stochastic and predictable only within ranges of probabilities.

The appreciation of seasonality as a significant source of human variability has a deep basis within the history of our discipline. In 1810, Samuel Stanhope Smith, an American clergyman, published a landmark volume on the effects of the environment on human biological variability. In this work, Smith noted that:

> The power of the climate to change the complexion is demonstrated by facts which constantly occur to our observation. In the summer season we perceive that the intensity of the sun's rays in our climate tends to darken the colour of the skin, especially in the labouring poor who are more constantly than others, exposed to their action. In the winter, on the other hand, the cold and keen winds which then prevail contribute to chafe the countenance, and to excite in it a sanguine and ruddy complexion. In the temperate zone, the causes of these alternate and opposite effects serve, in a degree, to correct one another . . . .. These effects, in countries where heat and cold succeed each other in nearly equal proportions, are transient and interchangeable (Smith, 1810, p. 43).

In more recent years, the seasonality of human biological and behav-

Table 2.1. *Phenomena for which there is evidence of seasonality in humans*

| | |
|---|---|
| Sudden infant death syndrome | Birth |
| Cleft palate | Mortality |
| Spina bifida | Abortion |
| Stroke | Breast feeding |
| Peptic ulcer | Growth |
| Asthma | Menarche |
| Hyperthyroidism | Hormone secretion (primates) |
| Schistosomiasis | Cellular immune function |
| Gastro-intestinal viruses | Diet |
| Hepatitis E | Nutritional status |
| Acute respiratory infections | Energy expenditure |
| Diarrhoea | Coital frequency |
| Schizophrenia | Loss of virginity |
| Depression | Autism |
| Suicide | |

ioural traits has been the subject of widespread investigation, with published papers spanning a range of features that encompass aspects of the physical and sociocultural environment (de Garine & Koppert, 1990), traits and behaviours of individuals (Stoll *et al.*, 1990), and population parameters (Goad *et al.*, 1976).

Table 2.1 presents a list of 29 phenomena for which there is evidence of a relationship to seasonal cycles or to month of the year. While many fall within the range of normal variation, the majority are significant with respect to the health status and adaptive success of individuals and groups. Seasonality is associated with infectious disease (Ansari *et al.*, 1991; Zhuang *et al.*, 1991), with congenital defects (Goad *et al.*, 1976), with psychoses (Watson, 1990), with nutritional status (Novotny, 1987), and with the course of growth and maturation (Sanders, 1934). Though not listed here, ritual, social, and subsistence activities are no less responsive to, and organized around, the ebb and flow of environmental conditions that accompany seasonal change (e.g., Rappaport, 1968).

In fact, only a few hypotheses regarding seasonal associations that have been tested have been rejected. A review of the literature revealed such rejection only in the case of Down's syndrome and of Alzheimer's Disease, though there are some positive reports for the latter. Even accepting the bias against negative results in scientific publishing, it is clear that we as a species are enormously responsive to the seasonal changes and environmental cycles that characterize our ecosystems.

The ability to respond adaptively to a changing habitat is an important part of our heritage as mammals, those vertebrates that display significant buffering from the fluctuations of the environments in which they live. The evolution of a menstrual cycle in anthropoid females was an important step in unlocking the close relationship of breeding activity to season. And the

Table 2.2. *Effects of environmental cycles on sleep–wake cycles and temperature rhythms in one human subject*

| Day | Condition | Observation |
|-----|-----------|-------------|
| 1–6 | Natural | Period lengths ≈ 24 hours. Cycles and rhythms in phase with day–night cycle |
| 7–23 | Isolated underground | Cycles and rhythms drifted toward a 26.5 hour period |
| 24+ | Natural | Cycles and rhythms synchronized to the 24-hour day |

From Takahashi & Zatz (1982).

implications of continuous sexual receptivity for the development of complex social relationships have been touched on by many writers.

As ancestral hominoids spread further from the arboreal habitats to which the earliest primates had been adapted, and in particular, as our hominid ancestors began to exploit a wider range of ecological niches, expanding into the temperate and eventually the sub-Arctic zones of the Old World, the need to develop mechanisms that would insure population survival in the face of regular and often severe changes in temperature, water, food sources, and infectious organisms became ever more pressing.

The phenotypic plasticity of *Homo sapiens* is surely traceable to the broad range of environments inhabited by populations of our species. But, in many aspects, it is just as certainly rooted within the seasonal fluctuations that characterize the majority of those environments. Seen in this way, environmental seasonality is not so much a stress to which contemporary human populations must adapt, but rather a basic component of the ecosystems within which our ancestors evolved, a component that has left its mark indelibly upon the makeup of our biology and our behaviour.

Much of human existence displays a periodicity, none more familiar than circadian rhythms: self-sustained oscillations in behavioural, physiological, and biological functions with periods approximating 24-hour cycles. In their review of the subject, Takahashi & Zatz (1982) noted that this periodicity is generated endogenously through oscillators that are synchronized to the daily environmental cycle of light and darkness. The sensitivity of these rhythms to exogenous factors has been demonstrated experimentally on a number of occasions. Table 2.2 summarizes the results of one experiment in which the sleep–wake cycles and body temperature rhythms of an adult male subject were altered by isolation from external stimuli and then restored to a normal cycle when contact with the world was re-established.

It is impossible to review all of the evidence of seasonality in the traits and environments of human populations. For this reason, the remainder of this chapter will deal with three cases from the literature. These cases have been chosen as examples of the significance of the interplay between human populations, their biology, and the seasonality of the environments within which they live and persist through time. These three are: 1. growth and development; 2. conception and birth; and 3. nutrition and disease.

### Growth and development

Seasonal periodicity in the patterns of human growth and development have been observed since Buffon noted such an effect in the growth of Count de Montbeillard's son. Formal research by human biologists and anthropologists dates back more than half a century (Bowditch, 1872; Nylin, 1929; Sanders, 1934). In his review of the literature, Bogin (1977) catalogued 29 studies of seasonal variation in rates of growth, noting that the studies point toward a 'marked seasonality in rates of growth for height and weight'. Studies that failed to reveal an effect tended to display deficiencies in methodology. In particular, Bogin notes a common flaw of too-infrequent a schedule of examinations to detect seasonal inflections.

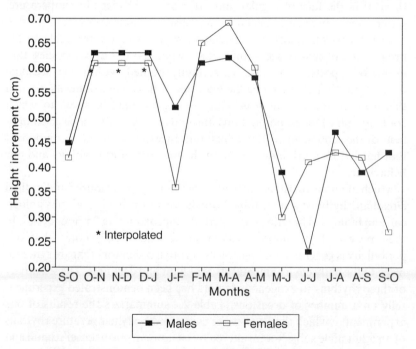

Fig. 2.1. Monthly height increments, for 5–7-year-old Guatemalan children (from Bogin, 1977).

Table 2.3. *Correlation between mean monthly increments and hours of insolation per month, Guatemalan subjects*

| Sex | Age (yr) | Height increment (cm/month) | Weight increment (kg/month) |
|-----|----------|------------------------------|------------------------------|
| M | 5.0–6.9 | 0.72** | −0.07 |
| F | 5.0−6.9 | 0.76** | 0.10 |
| M | 11.0–13.9 | 0.24 | −0.29 |
| F | 11.0–13.9 | 0.29 | −0.08 |
| M | 14.0–17.9 | 0.50* | −0.36* |
| F | 14.0–17.9 | 0.69** | −0.17 |

*$p < 0.05$, **$p < 0.001$.
From Bogin (1977).

Bogin studied the periodicity and seasonality of growth of children from the upper socioeconomic strata (SES) of the population of Guatemala City. Guatemala was chosen because of its lack of variation in daylight ($1\frac{1}{2}$ hours over the year) between summer and winter solstices, and upper SES children were sampled to minimize the effects of undernutrition and disease. Each child was measured monthly from September to October of the following year, except for November and December, the school holiday period.

Bogin's data indicate that the growth of his sample was associated with seasonal changes in the hours of insolation. During the period of his study, the difference in hours of sunlight between the months with the most and the least insolation was 57%. Fig. 2.1 shows the monthly increments in height for the children of his sample, with the greatest values associated with periods of greater sunlight.

Table 2.3 presents the correlations between the mean monthly increments of height and the mean hours of insolation per month during the period of his study. It can be seen that significance was found only for height and only in the ages before and after the adolescent spurt. Among children and older teenagers, increased growth in stature is associated with increased hours of sunlight, which Bogin attributed provisionally to the role of ultraviolet light in the synthesis of cholecalciferol, vitamin $D_3$.

It is also important to note that the seasonality of growth found by Bogin was detected in a population of upper socioeconomic status, without protein–energy malnutrition and buffered from the deleterious effects of the infectious diseases that can affect growth and that has a seasonal component. In other words, while the periodicity of growth may be intensified by seasonal fluctuations in an adverse environment, it also manifests itself under the less-harsh conditions experienced by the more affluent.

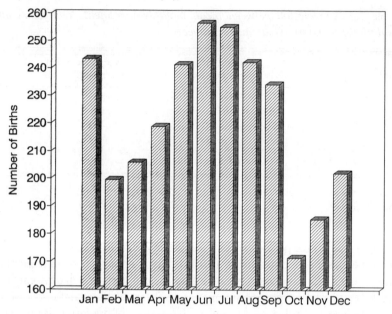

Fig. 2.2. Number of births by month, rural highland Lesotho (from Huss-Ashmore, 1988b).

### Conception and birth

The patterns of birth, and by extension, of conception, show a clear seasonal effect in most, though not all, of the studies in the literature. Fig. 2.2 presents the results reported by Huss-Ashmore (1988b) from research conducted in rural highland Lesotho. The data consist of the number of births per month, adjusted for variation in the number of days per month. The highest levels are seen in June and July, when birthrate is significantly higher than the expected, along with a short-term peak in January. Huss-Ashmore extrapolates these birth figures so as to derive two peaks in conception: one in April during the early harvest season and the other in September and October, during early Spring. Since factors limiting population growth are seen by her as being 'of greatest ecological interest', she notes further that the peaks are separated by two troughs, when conceptions fall below the expected frequency. Huss-Ashmore interprets these observations in economic terms, with climatic changes regulating the scheduling of agricultural tasks and migration, which affect coital frequency.

In general, birth seasonality is most marked among traditional farming societies, and least marked among populations of the world's industrialized nations. This may reflect the role of improved family planning in the latter

Table 2.4. *Frequency and extent of dietary shortages in traditional human societies*

| Frequency | n | % | Extent | n | % |
|---|---|---|---|---|---|
| Rare | 33 | 28.7 | Mild | 41 | 36.3 |
| Occasional | 28 | 24.3 | Moderate | 39 | 34.4 |
| Annual | 27 | 23.5 | Severe | 33 | 29.3 |
| Frequent | 27 | 23.5 | Not reported | 5 | |
| Not reported | 3 | | | | |

From Gaulin & Konner (1977).

instance (Halli, 1989) and the importance of ecological cycles in the former. At the same time, the existence of correlation of births with factors such as latitude (James, 1990) as well as the prevalence and intensity of disease (Bantje, 1987), suggest the operation of other determinants, both biological and behavioural.

### Nutrition and nutritional status

The third example discussed here involves nutritional status. The ability of the diet to satisfy the nutritional requirements of a population is dependent not only upon features such as habitat, technology, and biomass, but also upon the effects of fluctuations in the environment upon food availability. In their survey of the literature, Gaulin & Konner (1977) demonstrated that, while the proportion of energy derived from protein, carbohydrate, and fat in the diets of traditional human societies was in line with current nutritional recommendations, such groups were at risk of regular and significant shortages of food (Table 2.4). The term 'seasonal hunger' has been used to indicate recurrent periods when a society recognizes symptoms of hunger due to shortages of food. Often this occurs when extra physical effort is needed to get the new crop in the ground, a time when food supplies are in shortest supply (de Garine & Koppert, 1990).

Populations with limited technology for the storage and preservation of food, which are otherwise less well-buffered against nutritional stress, and which face significant burdens of infectious disease and parasite infestation, are particularly at risk of malnutrition. Among such groups, the additional pressure imposed by environmental seasonality becomes the trigger responsible for acute malnutrition, increased morbidity and a surge in the already-high rates of mortality.

Fig. 2.3 presents the adequacy in 1954 of the diets of rural families from Quezaltepeque, El Salvador in February, during the dry season, and in September, the time of rain (Flores & Flores, 1984). The data indicate a

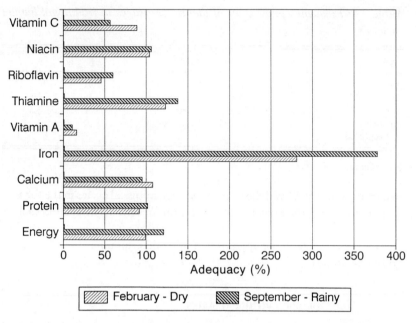

Fig. 2.3. Nutritive adequacy of family diets, Quetzaltpeque, El Salvador, 1954 (from Flores & Flores, 1984).

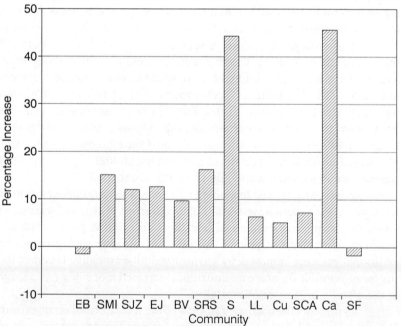

Fig. 2.4. Percentage increase in energy intake in the rainy season, for Guatemalan preschool children from several different communities (indicated by EB, SMI, etc.) (from Flores & Flores, 1984).

decrease of almost 25% in the adequacy of energy intake during the dry, compared to the rainy, season along with an overall decrease in the quality of the diet.

The effects of seasonality upon the diets of children may also be seen in Fig. 2.4, which presents the differences in energy intake in the rainy season, compared to the dry season, in preschool Guatemalan children from 12 communities. While a few of the 12 showed only a minor difference, others display variability ranging from 10% to over 40%.

Nutriture, or 'the state resulting from the balance between supply of nutrition on the one hand and the expenditure of the organism on the other' (McLaren, 1976), can be seriously compromised by disease, through the synergism of nutrition and infection (Scrimshaw *et al.*, 1966). The marked seasonality of the great bulk of the infectious diseases adds yet another dimension to any concern with nutritional status and season. However, it must be kept in mind that human populations have developed behavioural strategies for coping with periodic shifts in their environment. The result is that seasonal effects upon nutritional status might not be as great as expected, due to both metabolic and sociocultural adaptations, a conclusion reached by workers in Central America (Flores & Flores, 1984) and in Africa (de Garine & Koppert, 1990). In other words, human adaptability to seasonal limitations of food can be an important factor in the buffering of populations.

### Discussion

Three examples of seasonal effects upon human biology have been presented briefly in this chapter. The first dealt with growth and development, the second with conception and birth, and the third with food intake and nutriture. Taken together, they provide the basis for three statements about the human species, which are especially relevant for any consideration of seasonality and human biology.

First, *a wide range of human biological processes display periodicities that seem to exist apart from exogenous influences*. These periodicities operate at the cellular level and are regulated by neurophysiological processes mediated through patterns of hormone secretion and receptor sensitivity. They serve as part of the physiological regulation of homeostatic constancy and developmental stability. Their effects are seen, for example, in the menstrual cycle, in the control of body temperature, and in the regularity that characterizes human growth along with its spurts, shifts, and inflections. Periodic cycles are innate biological phenomena, which display their own patterns, yet which have been intricately synchronized during the course of evolution with the cycles of the surrounding environment.

Second, *periodic processes in humans are biobehavioural in nature*. In attempting to understand stability and change in human biological systems, and especially to analyze these phenomena within the context of seasonality, we must deal in realms that are both biological and behavioural, and particularly we must search for and analyze them as interactive, and not just as additive.

Third, *the study of seasonal components of human biological variability reveals the exquisite sensitivity of human groups to their ecosystems.* It is no accident that the preponderance of research into seasonality has adopted an ecological perspective as the framework that promises to provide the greatest depth of understanding. The environment exerts a powerful effect upon human populations, and its interactions with their gene pools are particularly important within the context of seasonal cycles.

In short, the study of seasonality within the boundaries of human biology is the study of the interrelations that exist between ourselves as individuals, as well as the populations within which we exist, and the ecosystems of which we are part. Both the human and the environmental components display repetitive cycles, directional inflections, and random shifts, and they come together within the network of biobehavioural interactions that characterize our species.

If the study of seasonality is so important, why then has so little systematic attention been paid to it? Why is it that a scan of the indexes of the textbooks and the major publications by human biologists fails almost without exception to reveal any mention of seasonality, or at best a few lines on wet versus dry, hot versus cold, or plenty versus famine? Or why is it, when a paper does focus upon seasonal fluctuations, more often than not, season becomes another stress, added to that long list of diseases, discomforts, and debilitations that are inflicted upon us. Huss-Ashmore (1988a) suggests that the 'continuing neglect' of seasonality has at least four causes: 1. it is too complex, requiring comprehensive interdisciplinary study; 2. it is too expensive, requiring more than single-visit surveys or short-term data-skims; 3. it is too uncomfortable, requiring the researcher to live under difficult conditions; and, 4. it is not seen as a problem, since the researchers themselves are residents of well-buffered industrialized and urbanized societies.

All of these are true. Yet, in order to incorporate what is undoubtedly a major force in population–environment interactions, analyses must move from simple univariate comparisons to more sophisticated multivariate approaches in which considerations of periodicity come into play. And models of stability and change must include variables reflecting seasonal change, in ways that go beyond the heuristic and descriptive to the operational and quantitative.

Perhaps, in the final analysis, seasonality is not something that human biologists study, but instead an attitude incorporated into their ways of thinking, and one that guides the research that is conducted and the designs that are formulated. Central to this attitude are the concepts of stability and change and the implications they have for the analysis of population–environment interactions. Population genetics has moved from the static to the dynamic, as has the analysis of human growth and development. Perhaps it is high time that human biologists take the next step and do the same for the study of human ecosystems.

### References

Ansari, S.A., Springthorpe, V.S. & Sattar, S. (1991). Survival and vehicular spread of human rotaviruses: possible relation to seasonality of outbreaks. *Review of Infectious Diseases*, **13**, 448–61.

Bantje, H. (1987). Seasonality of births and birthweights in Tanzania. *Social Science and Medicine*, **24**, 733–9.

Bloom, B.S. (1964). *Stability and Change in Human Characteristics*. New York: John Wiley.

Bogin, B.A. (1977). Periodic rhythm in the rates of growth in height and weight of children and its relation to season of the year. PhD Dissertation in Anthropology. Philadelphia: Temple University.

Bowditch, H.P. (1872). Comparative rates of growth in the two sexes. *Boston Medicine and Science Journal*, **10**, 434–41.

de Garine, I. & Koppert, S. (1990). Social adaptation to season and uncertainty in food supply. In *Diet and Disease in Traditional and Developing Societies*, ed. G.A. Harrison & J.C. Waterlow, pp. 240–89. Cambridge: Cambridge University Press.

Flores, M. & Flores, R. (1984). Effects of dependence on seasonally available food. In *Malnutrition: Determinants and Consequences*, ed. P.L. White & N. Selvey, pp. 207–19. New York: Liss.

Gaulin, S.J.C. & Konner, M. (1977). On the natural diet of primates, including humans. In *Nutrition and the Brain, Vol. 1*, ed. R.J. Wurtman & J. Wurtman, pp. 1–86. New York: Raven.

Goad, W.B., Robinson, A. & Puck, T.T. (1976). Incidence of aneuploidy in a human population. *American Journal of Medical Genetics*, **5**, 303–7.

Halli, S.S. (1989). The seasonality of births in Canada. *Journal of Biosocial Science*, **21**, 321–7.

Huss-Ashmore, R. (1988a). Introduction: why study seasons? In *Coping With Seasonal Constraints*, ed. R. Huss-Ashmore. MASCA Research Papers in Science and Archaeology, vol. 5. Philadelphia: The University Museum.

Huss-Ashmore, R. (1988b). Seasonal patterns of birth and conception in rural highland Lesotho. *Human Biology*, **60**, 493–506.

James, W.H. (1990). Seasonal variation in human births. *Journal of Biosocial Science*, **22**, 113–19.

McLaren, D.S. (1976). *Nutrition in the Community*. New York: John Wiley.

Novotny, R. (1987). Preschool child feeding, health, and nutritional status in Gualaceo, Ecuador. *Archivo Latinoamericano Nutricion*, **37**, 17–43.

Nylin, G. (1929). Periodical variations in growth, standard metabolism, and oxygen capacity of the blood in children. *Acta Medica Scandinavica*, Supplement 31.

Rappaport, R.A. (1968). *Pigs For the Ancestors: Ritual in the Ecology of a New Guinea People*. New Haven, CN: Yale University Press.

Sanders, B.S. (1934). *Environment and Growth*. Baltimore: Warwick and York.

Scrimshaw, N.S., Salomon, J.B., Birch, H.A. & Gordon, J.E. (1966). Studies of diarrheal diseases in Central America. VIII. Measles, diarrhea, and nutritional deficiency in rural Guatemala. *American Journal of Tropical Medicine and Hygiene*, **15**, 624–31.

Smith, S.S. (1810). *An Essay on the Causes of the Variety of Complexion and Figure in the Human Species*. New Brunswick, NJ: J. Simpson.

Stoll, C., Alembik, Y., Dott, B. & Roth, M.P. (1990). Epidemiology of Down syndrome in 118,265 consecutive births. *American Journal of Medical Genetics*, Supplement, **7**, 79–83.

Takahashi, J.S. & Zatz, M. (1982). Regulation of circadian rhythmicity. *Science*, **217**, 1104–11.

Watson, C.G. (1990). Schizophrenic birth seasonality and the age-incidence artifact. *Schizophrenia Bulletin*, **16**, 5–10.

Zhuang, H., Cao, X.Y., Liu, C.B. & Wang, G.M. (1991). Epidemiology of hepatitis E in China. *Gastroenterology Japonica*, **26**, 35–8.

# 3 *The influence of seasonality on hominid evolution*

R.A. FOLEY

## Introduction

From a palaeobiological perspective seasonality offers both a positive and a negative dimension. On the positive side is the fact that hominid evolution coincides with a marked increase in seasonal variation in climatic and environmental conditions, and hence is an obvious selective pressure. Furthermore, if seasonality is taken to be the occurrence of alternating periods of rich and poor resource availability, for whatever reason, then seasonality can be seen as a factor leading to more intense selection, and hence the focus for evolutionary change. On the negative side, though, is the difficulty of obtaining seasonal information from fossils and palaeoenvironmental contexts, such that it is difficult to link general patterns of seasonality to the adaptive responses that may develop. In this chapter I shall attempt to pursue a course that elaborates some of these positive elements without glossing too superficially over the negative ones. In particular, I want to look at some of the general mechanisms and environmental conditions that may have led to evolutionary responses on the part of hominids to seasonal conditions. By necessity this will focus more on aspects of foraging behaviour than on detailed physiological evidence.

## Climatic change, seasonality and hominid origins

There is a general consensus that the Hominidae (taken here to refer to the sister clade either of the other African apes generally or more specifically of *Pan*) have their origins in the late Miocene in Africa. Interpretations of molecular systematics in terms of a clock suggest a divergence from chimpanzees and/or gorillas at about 7.5 Ma (Holmes *et al.*, 1989). The earliest direct fossil evidence for hominids comes from localities in Kenya (Lothagam and Tabarin) dated to approximately 5.0 Ma (Hill & Ward, 1988). This material is very fragmentary, but by approximately 4.0 Ma

17

there is unambiguous evidence for hominids that are more bipedal than any living ape (Leakey, 1978). Thus the origins of the Hominidae as a clade lie in the late Miocene and distinct human-like adaptations have developed by the middle of the Pliocene.

The emergence of the hominids coincides with some important climatic and environmental changes. Evidence derived from marine cores shows that from the middle Miocene onwards there was a decline in global temperatures (Kennett, 1977). This decline accelerated markedly during the late Miocene, and culminated with the build up of the polar ice caps and the extension of continental glaciation during the Pleistocene. It is important to note here that the colder climates associated with the Pleistocene ice ages were but the end result of a much more long term trend towards reduced temperatures.

The impact of these climatic changes varied across the world. Of particular significance here is the effect upon the habitats of Africa, the location in which we find the earliest hominids. Bernor (1983) has described the middle Miocene habitats of Africa, and shown that at this time there was a broad and largely continuous distribution of evergreen, tropical forest. The climatic changes of the late Miocene led to the break up of these habitats. Woodland, bushland and grassland all expanded markedly during this period (Van Couvering & Van Couvering, 1976; Van Couvering, 1980). This process was significant for two reasons. The first was that it led to a much greater diversity of habitats, resulting in a mosaic pattern of variation, such as can be seen in parts of Africa today. This diversity and fragmentation, regardless of its detailed character, may have promoted evolutionary change and diversification through the formation of ecological boundaries and the isolation of populations. Kingdon (1980) and Gooder (1991) have used both refugia and vicariance models to account for the current distribution of living cercopithecoids as a response to this development of habitat structure.

The second factor was that the emerging habitats were far more seasonal. This can be measured both in terms of the increased seasonality of rainfall patterns and the availability of resources (Foley, 1987). During the middle Miocene it is probable that most regions of Africa received rainfall relatively evenly throughout the year. By the late Miocene and the Pliocene this was no longer the case, and, with the exception of the central and western parts of the continent, a four to eight month dry season became more typical. It is specifically in this context that hominids have their origins, as do most of the groups such as the baboons that have also adapted to primarily savannah conditions.

Hominid origins are therefore closely linked to the development of increasingly seasonal conditions in tropical Africa. This can be illustrated

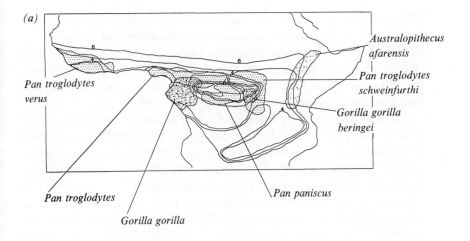

*(a)*

*Australopithecus
afarensis*

*Pan troglodytes
schweinfurthi*

*Pan troglodytes
verus*

*Gorilla gorilla
beringei*

*Pan troglodytes*

*Pan paniscus*

*Gorilla gorilla*

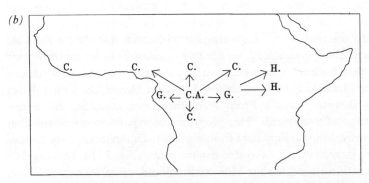

*(b)*

Fig. 3.1. (a) Distribution of African hominoids in relation to the number of dry
season months shown as contours. Distribution data from Wolfheim (1983), Lee *et
al.* (1988). (b) Proposed distribution of the common ancestor of the living African
hominoids, showing a core distribution for gorilla and increasingly peripheral ones
for chimpanzees and the earliest hominids. C = *Pan*, G = *Gorilla*, H = hominidae,
CA = common ancestor.

by looking at the distribution of living African hominoids, fossil hominids
and their habitats (Fig. 3.1a). The forested regions of Africa can be viewed
as an arc of distribution centred on the 'armpit' of Africa. Lowland gorillas
lie at the centre of distribution, with an island of eastward extension
consisting of the distribution of the mountain gorilla. Chimpanzee
distribution is much more extensive, and prior to human interference was
probably a larger arc surrounding and overlapping with the gorilla
distribution. The fossil hominids are wrapped around the eastern edge of
the chimpanzee distribution. Fig. 3.1a also shows how the distribution of
the living hominoids and the earliest hominid fossils relates to the number
of dry season months (defined as the number of months receiving less than

100 mm of rain per month). It can be seen that gorillas are confined to regions with no more than four months dry season per year. Chimpanzees can tolerate up to five months per year. In contrast, *A. afarensis* has been discovered in areas that at present have more than five dry months per year, and although it may have been wetter in the past, it is probable that this area was still relatively dry (Aronson & Taieb, 1981).

The suggestion may be made that a key difference between the early hominids and the other African hominoids is their ability to withstand more prolonged dry seasons. Indeed, one can perhaps generalise this further. It has been pointed out that hominoids were more successful during the warmer and wetter periods of the Miocene (Andrews, 1981). In contrast, cercopithecoids diversified more fully during the drier Late Miocene and Pliocene (Andrews, 1981). It has been suggested that this represents a greater tolerance of seasonal conditions. Their relative success within the hominoids may be viewed in a similar light. Humans and chimpanzees are the most widespread hominoids. They are also the most seasonally tolerant. The chimpanzee/gorilla/human split during the Late Miocene probably occurred from a common ancestor in the centre of their current distribution – Cameroon, Gabon, Zaire and the Central African Republic (the area lying between the Adamaou Massif, the Congo River and the Bongos Massif). From there they have expanded eastwards, westwards and southwards. This geographical expansion essentially consists of an evolutionary gradient from gorillas to chimpanzees to hominids, reflecting increasingly less forested environments (Fig. 3.1b). This suggests that speciation among the late Miocene hominoids of Africa may well have been parapatric (a view that is certainly consistent with the ambiguous molecular data on branching sequences), and driven by adaptation to increasingly seasonal conditions. In other words, the marked geographical patterning of African hominoids and the earliest hominids that has long been recognised (Kortlandt, 1984; Coppens, 1988–9) may represent a gradient of adaptive responses to the climatic changes of the Late Miocene, reflected geographically from west to east. Hominids were, in this perspective, merely the eastern extreme of the hominoid radiation (Coppens, 1988–9). No doubt because these populations faced the most extreme savannah conditions, they were also the ones that underwent the most marked evolutionary change.

## Seasonal influences on chimpanzee ecology

Given this apparent relationship between seasonality and the early divergence of the living African hominoids it is pertinent to examine the way in which seasonality influences chimpanzee behaviour and ecology. The premise here is not that early hominids were necessarily like chimpanzees, but that as they share the same basic problems of being large-bodied, highly social frugivores, then the patterns of response may have been similar. By looking not at a generalised model of chimpanzee behaviour but at the pattern of variation, it may be possible to highlight what elements of early hominid adaptation are most likely to have been influenced by seasonality.

A very simple model of seasonality is used here. If there is variation in resource availability during the course of an annual cycle, then it is usually possible to identify a period or periods of relatively abundant food supply (the rich season) and a period or periods of relative scarcity (the poor season). The question to be addressed is whether particular responses can be observed to seasonal variation in resource availability.

There are two extant species of chimpanzees – *Pan troglodytes* and *P. paniscus*. *P. paniscus* is confined to an area to the south of the river Congo and occupies areas with high rainfall and less marked seasonality. In contrast, *P. troglodytes* (the common chimpanzee) is far more widespread and occupies much drier and more seasonal habitats. For this reason these analyses will be confined to *P. troglodytes*.

Two potential and contrasting responses to periods of food scarcity can be proposed (Foley, 1987). One is that during periods of stress an animal will minimise energy expenditure. The second is that an animal will expend extra energy seeking out what few resources are available. The extent to which these two responses occur will be observable by looking at data on day range (the distance travelled by an individual during the course of the day's activities), time spent feeding, and feeding party size.

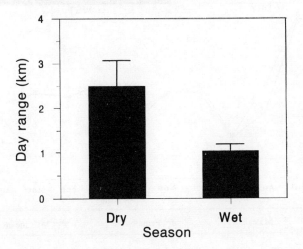

Fig. 3.2. Seasonal differences in day range length (in kilometres) among female chimpanzees at Gombe. Data from Goodall (1986).

## Day range

The long term study of the Gombe chimpanzees provides the most comprehensive data, including follows of specific individuals for several days during wet and dry seasons (Goodall, 1986). Fig. 3.2 shows the mean and standard deviation of day range length for females during wet and dry seasons. During the wet season day range length was 1.05 km (SD = 0.14, $N = 5$), whereas during the dry season it was 2.5 km (SD = 0.57, $N = 8$). The difference was statistically significant ($t = 4.95$, $p = 0.001$).

## Time spent feeding

As would be expected from the data on day range, time spent feeding by Gombe chimpanzees was also sensitive to seasonality. For both males and females the end of the wet season and the onset of the dry season was the period of maximal feeding (Fig. 3.3). Time spent feeding dropped during the course of the dry season and then climbed back up during the wet season. As can be seen from Fig. 3.4, which groups the monthly data by season, time spent feeding during the late wet/early dry season was approximately 60% (sexes pooled), compared to around 38% during the dry season when feeding time was at its lowest. At all times females spent longer feeding than males (by on average 3% more), but the differences between sexes were most marked during the dry season (6%) (Fig. 3.5).

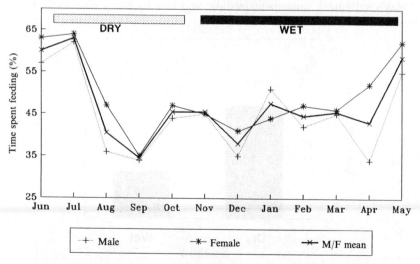

Fig. 3.3. The monthly variation in time spent feeding among Gombe chimpanzees. The seasonal distribution of rain is shown. Data from Goodall (1986).

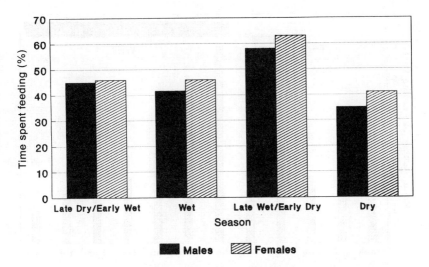

Fig. 3.4. Seasonality of time spent feeding among Gombe chimpanzees. The monthly data used in Fig. 3.3 is grouped here into the major seasons. See text for a discussion of the characteristics of each season. Data from Goodall (1986).

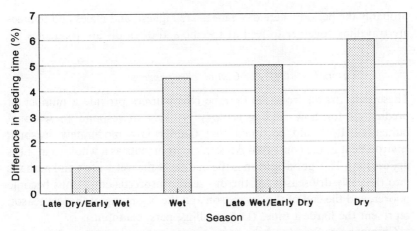

Fig. 3.5. Sex differences in time spent feeding by season for the Gombe chimpanzees. Females always spend more time feeding than males (data as for Fig. 3.4).

*Feeding party size*

Although chimpanzees live in relatively large stable communities, the size and make-up of feeding parties varies markedly. Fig. 3.6 shows the variation in feeding party size for the Gombe chimpanzees in relation to the seasons. As can be seen, the largest party size is found during the late dry/early wet season, when feeding time varies least between sexes. This is

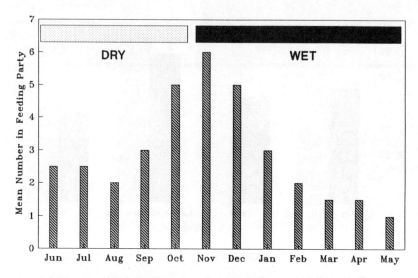

Fig. 3.6. Variation in feeding party size among Gombe chimpanzees. Note that the largest party sizes occur when the differences between sexes in feeding time is minimal (see Fig. 3.5). Data from Goodall (1986).

probably the period where day ranges are highest, and males and females are travelling together to feed at specific and often distant food sites.

### Seasonal strategies of Gombe chimpanzees

These data drawn from the Gombe chimpanzees provide a number of predictions for how feeding ecology is likely to respond to seasonal influences. It should be noted that Gombe is a moderately seasonal environment in the context of African environments as a whole (5 months dry season). The 'good' season is the middle of the wet season and perhaps into the early dry season; as the dry season proceeds conditions become poorer, and the end of the dry season and the beginning of the wet season represent the hardest times (D.A. Collins, pers. commun.).

Chimpanzees foraging behaviour is sensitive to this pattern of resource availability. As the wet season comes to an end day range increases, feeding party size reduces, and the differences between males and females become more marked. With the onset of dry conditions feeding time increases, and then decreases, followed by an increase in feeding party size at the end of the dry season and the beginning of the wet.

The Gombe data seem to suggest that in terms of the two responses outlined above – reducing energy or seeking out remote resources – the chimpanzee strategy is a mixed one. When conditions begin to change additional energy is expended in the acquisition of food from a larger area,

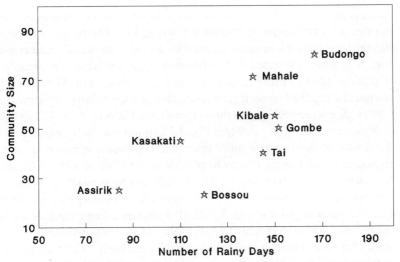

Fig. 3.7. The effect of increasing seasonality on chimpanzee community size. As rainfall becomes increasingly seasonal the size of chimpanzee communities decreases. 'Community' here refers to the larger units of socially interacting chimpanzees; these are relatively closed units, especially for males, but there is considerable fission and fusion within them on a daily basis for feeding and other activities. Data on rainfall for Budongo, Gombe, Assirik, Kasakati and Mahale come from McGrew *et al.* (1981). Data for Kibale, Bossou and Tai Forest are estimated from Walsh's (1981) seasonality indices. Data on chimpanzee community size for Mahale, Gombe, Bossou, Budongo, Assirik, Kibale and Kasakati (including Filabanga) come from Hiraiwa-Hasegawa *et al.* (1984, Table 6) and references therein, and Wrangham (1986). The community size for the Tai Forest is from Boesch and Boesch (1981). Where several communities are provided the median is used.

particularly by females, but that as dry conditions become more prolonged they switch to a reduction in energy expenditure and reduce the time spent feeding. This fits the general predictions made by Lee (1991). In terms of general principles that can be applied to the earliest hominids as they moved into more seasonal habitats we would expect that this variation would be enhanced, that day ranges would be larger, and that at the cusps of the wet/dry seasons there would be larger feeding groups spending considerably more time feeding.

### Comparative chimpanzee ecology

It may be argued that the Gombe chimpanzees are atypical, and it is certainly not the point here to use them as a simple analogue model for the early hominids. Rather the aim is to examine the trends that occur with seasonality in order to see what parameters might have changed in hominid evolution as the hominids began to inhabit more seasonal environments.

Some perspective can be gained on this by looking at the patterns across a number of chimpanzee communities living in different environments. Regrettably, data are available from only a small number of communities. Fig. 3.7 shows the relationship between a parameter of seasonality – number of days of rain per annum – and community size. The data were compiled from the literature; rainfall data is primarily from McGrew *et al.* (1981), supplemented by general meteorological information. The data on chimpanzee community size (see Fig. 3.7 caption for details of sources of data) uses medians where more than one community is reported, and it should be noted that there may be considerable variation within a locality; furthermore, some populations may be affected by pressure from human communities. For these reasons the data are not susceptible to further quantitative analysis, but Fig. 3.7 clearly indicates a trend towards smaller community size with increasing seasonality of rainfall. Although a whole suite of climatic and environmental variables are likely to interact to affect socioecology, nonetheless it is interesting to note that as chimpanzees move into environments that pose more extreme problems of seasonality, they appear to be forced to live in smaller communities.

## Evolutionary ecology of early hominids in seasonal environments

The expectations for the effect that seasonality would have had on early hominids are largely in terms of foraging behaviour and its socioecological context. However, not all seasonal environments present the same adaptive problems. This variation may help account for the fact that hominids in Africa between 3 and 1 My ago did not display a single evolutionary trend, but rather underwent an adaptive radiation.

### Early hominid community size

Extrapolating from living primates to extinct hominids must be done with caution. In this case it is not chimpanzees themselves that are being used, but the pattern of variation in relation to a seasonal parameter. If we apply the pattern observable in Fig. 3.7 to the early hominids, then the expectation would be that as hominids moved into more seasonal environments they would undergo a reduction in community size (*contra* Foley, 1989). This may well have been the case, and has major implications for reconstructing early hominid social organisation. However, a confounding variable is predation. The pattern of variation observable among chimpanzees is a response to resource structure changes. These populations, though, are not under any predatory pressure. The more open savannah environments in which we find the early hominids are likely to have many more large predators, and if, as Dunbar (1988) has argued, predator pressure drives group size upwards, then the trend observable in chimpanzee populations may have been reversed. Indeed, it may be the conflicting pressures of resource structure and predator pressure in highly seasonal environments that led to the formation of novel social structures among hominids.

### Pliocene hominid diversity

The precise number of taxa represented in the Pliocene and early Pleistocene fossil hominid record is a matter of dispute. There is general consensus though that there is more than one lineage present after 2.5 My ago (Klein, 1989). One lineage is that represented by *Homo*, characterised by relatively unspecialised dentition and greater brain size. The other, that of the 'robust' australopithecines (or *Paranthropus*) is usually described as megadontic, with very large, flat and worn molars, molarised premolars, and reduced incisors and canines. It has generally been argued that this latter lineage is dietarily specialised on coarse, hard and generally small-sized plant foods; what might be thought of as low quality herbivory involving high levels of processing time (see Grine (1989) for a recent review). In contrast, *Homo*, on the basis of both gross and micro dental anatomy, appears not to have been significantly different from chimpanzees, although there is archaeological evidence for a substantial increase in the amount of meat consumed (Bunn & Kroll, 1986; Potts, 1989). Given that there does not appear to be a significant gross habitat difference between these two lineages (White, 1989), it is pertinent to ask whether there is a more subtle seasonal effect.

### Seasonal responses and evolutionary diversity

The general pattern of resource availability in the drier regions of eastern Africa is that plant foods are relatively more abundant in the wet season, becoming increasingly scarce during the dry season (Foley, 1987). This imposes major constraints on herbivores, the bulk of which are the medium sized ungulates. These animals disperse widely during the wet season to exploit ephemeral food supplies as they are available. As the dry season develops they congregate around the permanent water supplies with the remaining stands of plant food. For a herbivore, therefore, the wet season is the 'good' season, and they live at low densities; the dry season is the 'poor' season and they live at very high, concentrated densities. The inverse is true of the carnivores. During the wet season their food supply is dispersed and they must spend a greater amount of time searching for it. In contrast, during the dry season prey items are more easily found in a few localised areas.

It has been argued that this is the basis for the divergence of the hominid lineage (Foley, 1987) during the Pliocene. In those areas where the medium and large sized ungulates were concentrated a principal means of ameliorating the effect of reduced plant food availability would be to exploit the animal resources, either through hunting or scavenging. Animal protein would have provided an extremely high quality resource that would even out the variation in annual food supply. A more omnivorous animal could switch back and forwards between meat in dry seasons and high quality plant foods in wet seasons. Given the relationship between encephalisation and dietary quality among primates (Milton, 1988; Foley & Lee, 1991), this may be the basis for the lineage leading to *Homo*, and the adaptive foundation for the other changes in life history that occur during the course of the evolutionary history of this genus (Foley, 1987; Foley & Lee, 1991).

However, not all East African environments are characterised by this seasonal pattern. Only in those areas where the large mammals were concentrated during the dry season would this be possible. In some regions, though, migration rather than concentration is the adaptive response. The classic example is that of the Serengeti ecosystem, where the ungulates exploit the regional differences in rainfall distribution by population movement. As rainfall decreases in the Serengeti it increases to the north, and so the herds migrate to these areas in southern Kenya (Maddock, 1979). Under these conditions a switch to meat would not be viable. It is more probable that in areas where this pattern predominated the only survival strategy in the dry season would have been to exploit the remaining, and generally low quality, plant foods. The characteristics of the robust australopithecines are consistent with this response to seasonal conditions.

The important conclusion to draw is that even when looking at long-term evolutionary patterns local conditions are of immense import-ance in determining how animals, including hominids, would respond to seasonal conditions. There was no single hominid strategy. Instead, the mosaic of East African habitats would have led to a patchwork of populations with distinctive adaptive characteristics (Fig. 3.8). This is certainly the case at the species level, and in all probability occurred at the sub-specific level as well. The implication is that it would be wrong to suggest that there is a simple or linear pattern of response to seasonality in hominid evolution.

It has been argued (Sinclair *et al.*, 1986) that it was the following of the large herds on their migratory routes that lay at the basis of the hominid

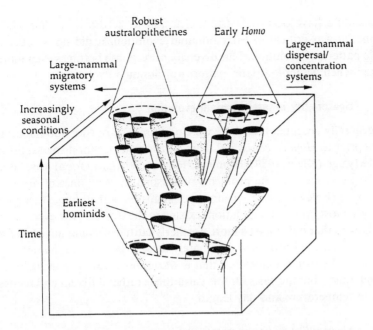

Fig. 3.8. Model of early hominid evolutionary divergence in relation to environmental conditions. The separation of the hominids is a response to local seasonal conditions and resource availability and would produce parallel adaptations (from Foley, 1987).

tendency towards greater meat eating. However, this is improbable in the context of early African hominids. The herds travel very fast; it is an adaptation only possible with highly precocial young such as is found among wildebeest. Primates, in contrast, have very altricial young, and the burdens of travelling large distances at speed while carrying infants are likely to have inhibited this as an adaptive response.

### Seasonality and early hominid evolution in Africa

It can be argued that specific seasonal conditions led to the appearance and subsequent diversification of the early African hominids. The early hominids were essentially those African apes that spread into the drier and more seasonal environments of the eastern part of the continent as these conditions became more pronounced. Adaptation to these conditions was associated with some significant changes in foraging patterns, particularly time spent feeding and travelling and the size and structure of foraging groups and communities. The variation in the specific resource structure of these more seasonal environments would have led to a diversity of responses and so a number of lineages rather than a single evolutionary trend should have occurred. The conclusion to be drawn is that while seasonality was critical to early hominid evolution, it did not impose a simple pattern of evolution. This diversity of response becomes even more marked when we look at later patterns of hominid evolution.

## Seasonality in later hominid evolution

It is generally accepted that there were no hominids outside sub-Saharan Africa prior to *Homo erectus*, and there is little evidence for dispersal prior to 1.0 My ago (Klein, 1989). Hominids have been tropical for at least 80% of their existence, and the problems of seasonality were ones of dry/wet oscillations (Foley, 1987). Even after the expansion of their range to higher latitudes, most hominid populations existed in the tropics. However, those populations that did colonise more northerly latitudes would have faced a new set of seasonal problems. Ultimately these problems are the same as those faced by tropical animals – that is, variation in resource availability through time – but the basis for this variation is rather different and relates more to temperature and day length.

### Seasonality and temperature

Clearly there is a general decline in mean annual temperature with latitude, and this will have had a major effect on hominid adaptation, as has been discussed extensively for the Neanderthals of the late Pleistocene (Trinkaus, 1981; Stringer, 1984). Specifically seasonal effects, as distinct from the general problem of reduced temperatures, are hard to pin down. However, one measure of seasonality to be considered might be the range of temperatures that must be withstood across an annual cycle (Table 3.1). In considering the pattern of temperature distribution it should be borne in mind that this has not remained stable during the Pleistocene but has fluctuated with the glacial cycles. These fluctuations would have steepened the temperature gradients at some periods. However, it is equally important to note that most of the Pleistocene did *not* consist of full glacial conditions (Shackleton *et al.*, 1988).

Until 1.0 My ago hominids were confined to less than 30° latitude north and south, and therefore experienced a thermal range of only 22 °C with a minimum temperature of 11 °C. Extension to 40° latitude would not have increased the range of temperature, although the minimum would have been down to 5 °C. This occurred with *Homo erectus* in Asia and persumably with *H. erectus* in Europe after 1.0 My ago, assuming current thermal conditions. This is likely as all the known occurrences are in inter-glacial conditions. Until the late Pleistocene, hominids do not appear to have been able to cope with the more extreme conditions, and were most

Table 3.1. *Mean maximal and minimal temperatures and thermal range by latitude*

| Latitude | Maximum temperature (°C) | SD | Minimum temperature (°C) | SD | Thermal range (°C) |
|---|---|---|---|---|---|
| 0–9 | 31.8 | 2.7 | 21 | 2.9 | 10.80 |
| 10–19 | 38 | 2.9 | 20.8 | 6.5 | 17.20 |
| 20–29 | 35 | 5 | 11 | 5.2 | 24.00 |
| 30–39 | 27.5 | 3.4 | 5.02 | 6.2 | 22.48 |
| 40–49 | 24.9 | 5.4 | −3.1 | 6.5 | 28.00 |
| 50–59 | 21.14 | 3.8 | −13.78 | 12.76 | 34.92 |
| 60–69 | 17.8 | 2.1 | −39 | 12.1 | 56.80 |

From Eyre (1968).

probably unable to withstand prolonged seasonal periods below freezing point. Gamble (1986) has argued that until the late Pleistocene hominid occupation of these latitudes was intermittent, and subject to local extinction whenever there was climatic deterioration. Thermoregulatory stress caused by exposure to extreme winter seasons is likely to have been a major barrier to hominid expansion, and underpins the idea that until the later Pleistocene hominids were essentially tropical animals, as indeed are all primates, with only very peripheral and tenuous dispersal beyond the warmer middle latitudes.

The exception to this pattern are the Neanderthals of Europe, which evolved after about 200 Ky ago during the penultimate glacial period, and which show permanent occupation of extremely cold conditions (equivalent to latitudes 50–60° today) during the last glaciation. It is now generally accepted that their facial prognathism, nasal size, robusticity and extreme limb proportions reflect adaptation to cold conditions. This is in marked contrast to the appearance of anatomically modern *Homo sapiens* in Europe and Eastern Asia between 40 and 30 Ky ago. These hominids show tropical limb proportions (Trinkaus, 1981), and the suggestion may be made that it was only with these populations that behavioural mechanisms developed that reduced the need for major anatomical and physiological adaptations to extreme thermal stress. This view would certainly be supported by the extensive evidence for marked differences in behaviour between modern and archaic hominids in Europe (Mellars, 1989).

Overall it may be suggested that four phases of response to thermal stress can be recognised in hominid evolution: 1. the Pliocene and early Pleistocene when hominids were restricted to tropical and sub-tropical thermal regimes, and indeed, where it can be argued that there was major selection for the ability to withstand considerable heat load (Wheeler,

1991a, b); 2. the middle Pleistocene, when hominids colonised the warmer mid-latitudes during times of relative warmth, but most probably became locally extinct with the onset of colder climatic conditions; 3. the later middle Pleistocene when local populations of archaic hominids adapted to extreme conditions of cold through major morphological changes; and 4. the later Pleistocene (after 50 Ky ago), when modern humans managed to colonise all parts of the world, independent of climatic conditions and without major morphological change. The principal implication of this view would be that the physiological adaptations to cold conditions that we find in modern humans are, in evolutionary terms, very recent – certainly less than 50 Ky ago and in all probability less than 20 Ky ago.

### Seasonality and time stress

While the cold temperature of a middle or high latitude winter may have been a barrier to hominid occupation, it is by no means the only one. Indeed it may be argued that while temperature may be the ultimate evolutionary constraint it operates through more proximate ones relating to resource structure and availability, and indeed there may be other factors that are independent of temperature, such as the competitive structure of the mammalian community (Turner, 1992).

Recent work (Dunbar, in press) in behavioural ecology suggests that time budgets may be one such proximate mechanism. Successful animals must have a positive energy budget, and the time available for foraging activities can be critical to this. Time is finite, and the more time spent acquiring food must mean less time for other activities. Colonisation of high latitudes by an essentially seasonal tropical species introduces a new set of time budgeting problems that are primarily seasonal.

As we saw above (Fig. 3.3), chimpanzees spend between 35% and 65% of their time engaged in feeding. These figures are best treated as approximate given the differences in field methods, especially with regard to the distinction between feeding time and time spent travelling between food patches. Assuming that in broad terms a chimpanzee spends about 50% of the day feeding – i.e. about 6 hours per day – then any reduction in daylength must result in a loss of time available for other activities if feeding levels are to be maintained. With increasing latitude, daylength is severely reduced for large parts of the year. At 30° the midwinter daylength is approximately 10 hours; at 40° it is 8.5 hours. Thus at 40° latitude to have maintained 6 hours foraging time would have taken up 70% of the available hours of daylight. Time stress would have been a primary difficulty for high latitude hominid foragers. The effect may have been particularly marked when it is borne in mind, as was discussed above, that increasing, rather than decreasing, foraging time is one of the strategies available to primates for avoiding seasonal stress.

If the time available for foraging (and other activities) is the critical seasonal problem facing hominids living in higher latitudes, then the question that follows is what coping strategies are available if the 'normal' primate one of increasing time spent acquiring food is not possible. One possible option is that seen in chimpanzees towards the end of the dry season, that of reducing energy expenditure. How feasible such a strategy would be under conditions of extreme thermal stress is extremely problematic, but it should certainly be considered that archaic populations such as the Neanderthals may have had some physiological mechanism that enabled them to survive longer periods on relatively reduced levels of nutrients. If this were the case then it would open up possibilities of exploring how Neanderthal physiology may have differed markedly from that of modern humans.

The second option is related to technology. Torrence (1983) has shown that the complexity of hunter–gatherer technology increases with latitude. She has argued that the key function of technology is that it will decrease the amount of time it takes to process food. Dependence upon technology is therefore a characteristic we would expect to see more highly developed in high latitude populations, which are faced with time-constrained seasonal stress. It is this factor that may underlie the contrast in patterns of technological development in Europe and Northern Asia compared to low latitude regions, and it could be argued that it is technology, including that associated with storage (Binford, 1980), that enabled modern humans to colonise northerly latitudes permanently without major anatomical adaptations and during a period of maximum glaciation.

### Conclusions

This chapter has set out to examine the overall effect of seasonal factors on the pattern of hominid evolution. It must be stressed at this point that while it is possible to identify seasonality in the past, it is far harder to show directly that this rather than other parameters may have been responsible for the course that hominid evolution has taken. It may well be that there are other confounding variables, particularly those such as temperature and rainfall, that are equally significant. However, with this caveat, it is possible to suggest that seasonality has been one of the major adaptive problems that hominids have faced during their evolution.

It has been argued that responses to seasonality have occurred in different ways at various times and with specific populations. It must be noted that hominid evolution is not a single seamless event, but a patchwork of such events. The diversity of responses is likely to have been much greater than has been suggested here. However, it can certainly be suggested that hominid origins are at least partly tied into the ability of an African ape to adapt to longer dry seasons ( > 5 months); that the diversity of early Pleistocene hominids in Africa reflects different responses to seasonal conditions; and that those later hominid populations that colonised higher latitudes would have faced entirely novel seasonal conditions relating specifically to temperature and daylength. The overall chronological span of the genus *Homo* suggests that these responses were markedly different between populations and varied in their success.

Although the data and analyses presented here are primarily relevant to the study of hominid evolution, it is also the case that this evolutionary background has implications for the way we approach the biology of living populations, the primary focus of this volume. The most important of these is that it is almost certainly the case that all modern humans today have a relatively recent tropical ancestor. While hominids may have lived in Europe and Asia for almost a million years, modern humans have only been there for less than 50 Ky. Such adaptations that we see to seasonal conditions in high latitudes must therefore be recent.

Secondly, as we are now coming to understand that *Homo sapiens* is but a small part of overall hominid diversity, we can recognise that there may well have been, among the archaic hominids, a very different set of responses to seasonality. In particular, the balance between behavioural, activity-based and physiological and anatomical responses may well have been different. Modern humans undoubtedly have their own distinctive, species-specific way of adapting to seasonal stress.

And finally, the point about human evolution is that it is not a ladder of progress leading to modern humans. Fossil hominids are best thought of as perfectly good species that happened to become extinct for the most part. As such they represent an immensely important comparative dataset for the study of the ways in which ecological strategies, for adaptation to seasonal problems as well as other adaptive situations, are developed. By incorporating living primates and extinct hominids into the study of human biology we are able to construct a much broader comparative framework, and thus identify what are the general principles underlying adaptation to seasonality. As a first step towards this, it may be suggested here that the relative patterns of time and energy spent foraging will underlie what we see in living humans, extinct hominids, and living non-human primates.

### Acknowledgements

I thank P.C. Lee, M.M. Lahr and H. Eeley for comments on earlier drafts.

## References

Andrews, P.J. (1981). Species diversity and diet in apes and monkeys during the Miocene. In *Aspects of Human Evolution*, ed. C.B. Stringer, pp. 63–98. Symposia for the Study of Human Biology No. 12. Cambridge: Cambridge University Press.

Aronson, J. & Taieb, M. (1981). Geology and paleogeography of the Hadar hominid site, Ethiopia. In *Hominid Sites: Their Geologic Settings. AAAS Selected Symposium*, ed. G. Rapp & C.F. Vondra, pp. 165–95. Washington: American Association for the Advancement of Science.

Bernor, R.L. (1983). Geochronology and zoogeographic relationships of the Miocene Hominoidea. In *New Interpretations of Ape and Human Ancestry*, ed. R.L. Ciochon & R.S. Corruccini, pp. 149–64. New York: Plenum.

Binford, L.R. (1980). Willow smoke and dogs' tails: hunter-gatherer settlement systems and archaeological formation processes. *Journal of Anthropological Research*, **35**, 255–73.

Boesch, C. & Boesch, H. (1981). Sex differences in the use of natural hammers by wild chimpanzees: a preliminary report. *Journal of Human Evolution*, **10**, 585–93.

Bunn, H.T. & E. Kroll (1986). Systematic butchery by Plio-Pleistocene hominids at Olduvai Gorge, Tanzania. *Current Anthropology*, **27**, 341–52.

Coppens, Y. (1988–9). Hominid evolution and the evolution of the environment. *Ossa*, **14**, 157–64.

Dunbar, R.I.M. (1988). *Primate Social Systems*. London: Croom Helm.

Dunbar, R.I.M. (in press). Time: a hidden constraint on the behavioural ecology of baboons. *Behavioural Ecology and Sociobiology*.

Eyre, S.R. (1968). *Vegetation and Soils: a World Picture*. London: Arnold.

Foley, R.A. (1987). *Another Unique Species: Patterns in Human Evolutionary Ecology*. Harlow: Longman.

Foley, R.A. (1989). The evolution of hominid social behaviour. In *Comparative Sociology*, ed. V. Standen and R.A. Foley, pp. 473–94. Oxford: Blackwell Scientific Publications.

Foley, R.A. & Lee, P.C. (1991). Ecology and energetics of encephalisation in hominid evolution. *Philosophical Transactions of the Royal Society of London, Series B*, **334**, 223–32.

Gamble, C. (1986). *The Palaeolithic Settlement of Europe*. Cambridge: Cambridge University Press.

Goodall, J. (1986). *The Chimpanzees of Gombe*. Cambridge, MA: The Belnap Press of Harvard University Press.

Gooder, S.J. (1991). A phylogenetic and vicariance analysis of some African forest mammals. PhD Thesis, University of Liverpool.

Grine, F.E. (ed.) (1989). *The Evolutionary History of the 'Robust' Australopithecines*. Chicago: Aldine de Gruyter.

Hill, A. & Ward, S. (1988). Origin of the Hominidae: the record of African large hominoid evolution between 14 My and 4 My. *Yearbook of Physical Anthropology*, **31**, 49–83.

Hiraiwa-Hasegawa, M. Hasegawa, T. & Nishida, T. (1984). Demographic study of a large-sized unit-group of chimpanzees in the Mahale Mountains, Tanzania: a preliminary report. *Primates*, **25**, 401–13.

36    R.A. Foley

Holmes, E.C., Pesole, G. & Saccone, C. (1989). Stochastic models of molecular evolution and the estimation of phylogeny and rates of nucleotide substitution in the hominoid primates. *Journal of Human Evolution*, **18**, 775–94.

Kennett, J.P. (1977). Cenozoic evolution of anatarctic glaciation, the circum-antarctic ocean, and their implications on global palaeooceanography. *Journal of Geophysical Research*, **82**, 3843–60.

Kingdon, J. (1980). The role of visual signals and face patterns in African forest monkeys of the genus *Cercopithecus*. *Transactions of the Zoological Society of London*, **34**, 431–75.

Klein, R.G. (1989). *The Human Career*. Chicago: Chicago University Press.

Kortlandt, A. (1984). Habitat richness, foraging range and diet in chimpanzees and some other primates. In *Food Acquisition and Processing in Primates*, ed. D.J. Chivers, B.A. Wood & A. Bilsborough, pp. 119–60. New York: Plenum Press.

Leakey, M.D. (1978). Pliocene footprints at Laetoli, Northern Tanzania. *Antiquity*, **52**, 133.

Lee, P.C. (1991). Adaptations to environmental change: an evolutionary perspective. In *Primate Responses to Environmental Change*, ed. H.O. Box, pp. 39–56. London: Chapman and Hall.

Lee, P.C., Thornback, J. & Bennett, E.L. (1988). *Threatened Primates of Africa: the IUCN Red Data Book*. Switzerland and Cambridge: IUCN.

Maddock, L. (1979). The 'migration' and grazing succession. In *Serengeti: Dynamics of an Ecosystem*, ed. A.R.E. Sinclair & M. Norton-Griffiths, pp. 104–209. London: Academic Press.

McGrew, W.C., Baldwin, P.J. & Tutin, C.E.G. (1981). Chimpanzees in a hot, dry, open habitat: Mount Assirik, Senegal, West Africa. *Journal of Human Evolution*, **10**, 227–44.

Mellars, P.A. (1989). Technological changes at the Middle-Upper Palaeolithic transition: economic, social and cognitive perspectives. In *The Human Revolution*, ed. P.A. Mellars & C.B. Stringer, pp. 338–65. Edinburgh: Edinburgh University Press.

Milton, K. (1988). Foraging behaviour and the evolution of primate intelligence. In *Machiavellian Intelligence*, ed. R.W. Byrne & A. Whiten, pp. 285–305. Oxford: Clarendon Press.

Potts, R. (1989). *Early Hominid Activities at Olduvai*. Chicago: Aldine de Gruyter.

Shackleton, N., Imbrie, F. & Pisias, N. (1988). The evolution of oceanic oxygen-isotope variability in the North Atlantic over the past three million years. *Philosophical Transactions of the Royal Society of London, Series B*, **318**, 679–88.

Sinclair, A.R.E., Leakey, M.D. & Norton-Griffiths, N. (1986). Migration and hominid bipedalism. *Nature*, **324**, 307–8.

Stringer, C.B. (1984). Human evolution and biological adaptation in the Pleistocene. In *Hominid Evolution and Community Ecology: Prehistoric Human Adaptation in Biological Perspective*, ed. R.A. Foley, pp. 55–84. London: Academic Press.

Torrence, R. (1983). Time budgeting and hunter-gatherer technology: In *Hunter–Gatherer Economy in Prehistory: A European Perspective*, ed. G.N. Bailey, pp. 11–22. Cambridge: Cambridge University Press.

Trinkaus, E. (1981). Neandertal limb proportions and cold adaptation. In *Aspects of Human Evolution*, ed. C.B. Stringer, pp. 187–224. Symposia of the Society for the Study of Human Biology No. 21. London: Taylor & Francis.

Turner, A. (1992). Large carnivores and earliest European hominids: changing determinants of resource availability during the Lower and Middle Pleistocene. *Journal of Human Evolution*, **22**, 109–26.

Van Couvering, J.A.H. (1980). Community evolution in East Africa during the Late Cenozoic. In *Fossils in the Making*, ed. A.K. Behrensmeyer & A.P. Hill, pp. 272–97. Chicago: Chicago University Press.

Van Couvering, J.A.H. & Van Couvering, J.A. (1976). Early Miocene mammal fossils from East Africa: aspects of geology, faunistics and palaeoecology. In *Human Origins: Louis Leakey and the East African Evidence*, ed. G. Isaac & E. McGown, pp. 155–207. Menlo Park, CA: W.A. Benjamin.

Walsh, R.P.D. (1981). The nature of climatic seasonality. In *Seasonal Dimensions to Rural Poverty*, ed. R. Chambers, R. Longhurst & A. Pacey, pp. 11–12. London: Francis Pinter.

Wheeler, P.E. (1991a). The thermoregulatory advantages of hominid bipedalism in open equatorial environments: the contribution of increased convective heat loss and cutaneous evaporative cooling. *Journal of Human Evolution*, **21**, 107–16.

Wheeler, P.E. (1991b). The influence of bipedalism on the energy and water budgets of early hominids. *Journal of Human Evolution*, **21**, 117–36.

White, T.D. (1989). The comparative biology of 'robust' australopithecines: clues from context. In *The Evolutionary History of the 'Robust' Australopithecines*, ed. F.E. Grine, pp. 449–84. Chicago: Chicago University Press.

Wolfheim, J.H. (1983). *Primates of the World: Distribution, Abundance and Conservation*. Cher, Switzerland: Harwood Academic Press.

Wrangham, R.W. (1986). Ecology and social relationships of two species of chimpanzees. In *Ecological Aspects of Social Relationships in Birds and Mammals*, ed. D.I. Rubenstein & R.W. Wrangham, pp. 352–78. Princeton: Princeton University Press.

# 4 Environmental temperature and physiological function

M.A. STROUD

## Introduction

Worldwide, humans are exposed to tremendously varied environmental conditions, which include temperature ranges of $-80\,°C$ to $+55\,°C$, relative humidities of less than 2% to 100%, windspeeds in excess of 180 kph and radiative loads of greater than $1000\,Wm^{-2}$. Yet, despite these variations, core temperature is usually maintained within circadian limits of approximately $36\,°C$ to $38\,°C$ at rest, rising to about $39\,°C$ during heavy exertion. This thermoregulation is achieved using a combination of behavioural strategems and physiological changes which respond to both conscious assessments of the environment and sub-conscious temperature information supplied by both peripheral and central thermo-receptors to a hypothalamic regulatory centre. The behavioural and physiological responses act to alter components of thermal equilibrium as expressed in the heat balance equation:

$$S = M \pm E \pm R \pm C \pm K$$

where $S$ is heat storage, $M$ is metabolic heat production, $E$ is heat exchange by evaporation, $R$ is heat exchange by radiation, $C$ is heat exchange by convection, and $K$ is heat exchange by conduction.

These responses are effective over a wide range of environments, providing that clothing and activity levels are appropriate, but in extreme environmental conditions they can impose severe physiological stresses, which on occasions cannot be met. However, individuals vary considerably in their ability to resist such stress and much of this variation is influenced by physiological adaptations to the environment. These adaptations include: 1. evolutionary changes, which have arisen in races through natural selection; 2. acclimatization changes, which develop in individuals within periods that are quite short or, at most, their own lifetime; and 3. habituation changes, which simply reflect the decrease in physiological responses to a repetitive stimulus.

This chapter aims to give an overview of the physiological responses to hot and cold environments, including the shorter term adaptive changes that might be expected to alter on a seasonal basis.

## Physiological responses to hot environments

Human responses to the heat are well documented and are generally non-contentious. Behavioural stratagems include alterations in activity levels, body position and clothing, whilst the chief physiological changes involve alterations in peripheral vasomotor tone and sweating.

The vasomotor changes occur chiefly as a response to increases in core temperature and lead to active vasodilatation triggered by sympathetic dilatory fibres combined with the release of sympathetic adrenergic constrictor tone. However, the vasoactive agonist has not been identified and it is not clear to what extent the vasodilatation is mediated by the release from sweat glands of locally active compounds (Sawka & Wenger, 1988). Local skin temperature also has direct effects on vessel diameter. The peripheral vascular changes lead to an increase in surface blood flow and hence convective, conductive and radiative heat loss, although the response is only effective as long as ambient temperature does not exceed skin temperature. The changes also lead to a decrease in the effective insulation of the skin and subcutaneous tissues.

Sweating responses to the heat are also largely dictated by core temperature with increases in activity mediated by sympathetic cholinergic fibres. However, circulating catecholamines also lead to an increase in sweat production (Allan & Roddie, 1972). Once again, there are direct local effects, which include an increase in sweat secretion following the general increase in skin blood flow and inhibition of sweat gland function with increasing skin wettedness.

The changes in peripheral blood flow and sweating in response to heat are highly effective mechanisms unless high relative humidities limit the potential for evaporative heat loss. Lind (1963) demonstrated that over a wide range of environments known as the 'prescriptive zone', a given level of activity leads to the establishment of a steady state core temperature that is nearly independent of environmental temperature (Fig. 4.1). However, as illustrated in the figure, increasing environmental temperature eventually leads to an inflexion point at which, for a given work load, heat dissipation becomes inadequate. At this point, equilibrium core temperature begins to rise rapidly and at slightly higher environmental temperatures a steady state cannot be attained.

The specific thermoregulatory responses to heat are accompanied by changes reflecting physiological stress, for example an increase in heart rate

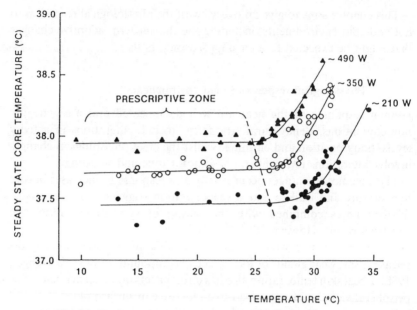

Fig. 4.1. Steady state core temperature at three metabolic rates and increasing environmental temperature (redrawn from Lind, 1963).

due to discomfort, high skin blood flow and peripheral blood pooling. However, under most circumstances, acclimatative changes in the primary responses to heat can rapidly improve tolerance and decrease the level of stress in a given environment.

## Physiological adaptation to hot environments

The most important adaptive changes to repeated heat exposure involve improvement in the peripheral vasomotor and sweating responses. The improvement in vasodilatation includes a decrease in the core temperature required for onset combined with greater increases in surface blood flow (Wenger, 1988). However, because of the effectiveness of the improvements in sweat rate (see below) and hence heat tolerance, overall skin blood flow in an acclimatised individual tends to be lower or unchanged relative to that in an unacclimatised person performing the same exercise in a given combination of exercise and environment.

The improvement in the sweating response is generally highly effective and also includes a decrease in the core temperature required for onset. This is combined with an increase in the maximal sweat rate from about 1.5 to 2–3 l/hr, although maximum daily production does not often exceed 15 litres (Wenger, 1988). Over the longer term, there are also changes in regional sweating distribution with increased sweating activity in areas that were previously of low capacity and, during more prolonged heat exposure, a limitation of the decline in sweat rate that is seen as heat exposure progresses (probably due to dehydration and/or hydromeiosis). Interestingly, however, long term dwellers in hot climates finally end up sweating less, appearing to limit sweat rates to match evaporation rates (Kuno, 1956). This is a change that permits the efficient use of water resources, although the mechanism through which this adjustment is made is unclear.

Other adaptive responses to heat include alterations in body fluid distribution accompanied by changes in both circulating electrolytes and their homeostatic hormones. The fluid changes include a temporary increase in plasma volume of up to 25%, returning towards normal within about 10 days (Senay *et al.*, 1976; Bass & Henschel, 1956). There is also a transient increase in extra-cellular fluid volume and an increase in the haemodilution change seen with exercise (Senay, 1979). These changes are accompanied by an increase in stroke volume and a decrease in heart rate although most of the heart rate changes reflect the overall decrease in thermal stress. The changes in electrolyte homeostasis include enhanced secretion of ACTH, cortisol, aldosterone and ADH leading to greater water and salt retention (Wenger, 1988). In particular, sweat sodium losses are reduced to less than $5 \, \text{meq} \, l^{-1}$ (Robinson & Robinson, 1954) and urinary sodium losses can be almost abolished after only a few days of repetitive heat stress.

Fig. 4.2 illustrates the effectiveness of acclimatisation in terms of rectal temperature, heart rate and sweat rate for fit young men working in the heat.

The timing of the acclimatisation responses has been the subject of many studies, with most concluding that they are near maximal within seven days if induced by the combination of physical work and a hot environment. However, there is no sharp end to the improvements seen and lesser degrees of acclimatisation are achieved more slowly by either resting from heat exposure or hard physical work under temperate conditions. Following the end of heat exposure, acclimatisation declines over 20 to 40 days.

In view of the relative rapidity of onset and decline of these adaptative changes to heat, it is not surprising that several studies have documented seasonal changes in acclimatative status corresponding to either partial or complete acclimatisation during the summer and the loss of that acclimatisation in winter. Such studies have included: a summer increase of up to 6 °C in the environmental temperature marking the upper end of the prescriptive zone at high work levels (Kuhlemeier *et al.*, 1977); improved responses to standard heat tests performed in the summer (Hori *et al.*, 1974); increases of 30–40% in summer circulating volume (Bazett *et al.*,

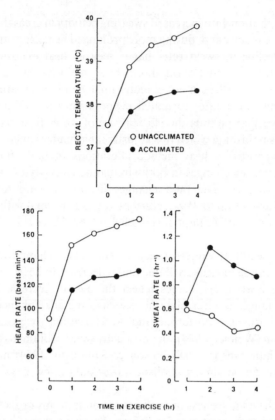

Fig. 4.2. The effectiveness of acclimatisation in 10 young males during 4 hours work in the heat (redrawn from Wyndham *et al.*, 1964).

1940); and improvements in tilt table tolerance (an experimental technique for evaluating cardiovascular responses to postural change) during the summer (Shvartz & Meyerstein, 1970).

## Physiological responses to cold environments

Human physiological responses to cold environments are less well documented than those to the heat and there are a number of areas of contention with regard to the relative importance and underlying mechanisms of some of the changes. The uncertainties probably reflect the fact that thermal homeostasis tends to be maintained by behavioural changes, including protection using clothing and shelter. Certainly, the effectiveness of such protection has been documented for men living through the winter in Antarctica, who were shown to be generally too warm rather than too cold (Norman, 1965). Changes in voluntary muscular activities can also increase heat production by up to 15-fold and some groups use seasonal migration to avoid excessive cold exposure. However, important physiological responses do occur and these include changes in peripheral vasoconstriction, shivering thermogenesis, and non-shivering thermogenesis (NST).

Peripheral vasoconstriction is increased with exposure to cold, a change that is mediated through a combination of a central increase in sympathetic, adrenergic tone in response to falling skin temperature. There is also a direct local effect of cold on small peripheral arterioles (Keatinge & Harman, 1980). The changes lead to a decrease in blood flow to the skin, minimising surface heat losses from conduction, convection and radiation and maximising the insulative value of both the skin and subcutaneous tissues.

The shivering response is also triggered by falls in skin temperature although, in addition, a fall in core temperature allows release of resting hypothalamic inhibition (Hemingway, 1963). Although the response is generally well documented, there remain some areas of debate regarding both its importance and its underlying mechanisms.

For example, in contrast to the heat production through voluntary muscular activity, shivering is only capable of increasing resting metabolic rate by about five fold and even that figure is exceptional. Its capacity significantly to influence thermal equilibrium has, therefore, been questioned and, from a teleological viewpoint, it has been suggested that the response has developed in order to maintain muscle plasticity while static or sleeping. However, although maintained plasticity would permit the efficient resumption of activity and hence active heat production, studies on the regional distribution of shivering demonstrate that truncal muscle groups are involved at an earlier stage of the response and shiver more intensively than limb muscles (Tikuisis *et al.*, 1991). This would favour the belief that shivering is designed to limit falls in core temperature and, clearly, even a doubling of resting heat production can have a significant effect on thermal balance in a reasonably insulated human.

Another area of current debate involves muscle substrate usage during shivering. Young *et al.* (1989) have demonstrated that nutritional carbohydrate deprivation and consequent depletion of resting muscle glycogen levels does not affect shivering capacity and argue that free fatty acids (FFAs) must, therefore, be the most important substrates. They support this argument by pointing out that even violent shivering only accounts for approximately 30% of maximal oxygen consumption and hence the situation is analagous to the predominant use of FFAs as substrates during low intensity exercise. Conversely, Martineau & Jacobs (1989) have shown that glycogen depletion can lead to a decline in shivering heat production and argue that since shivering does not involve all muscle units equally, glycogen would be expected to be the chief substrate in actively shivering units that may be working at high fractions of their maximal oxygen uptake.

In addition to shivering as a means of non-voluntary heat production, it is well established in small mammals that cold exposure can lead to an increase in NST. This NST relies upon the activation of futile metabolic cycles in brown adipose tissue mediated by the sympathetic nervous system and the release of nor-adrenaline (NA). Although present in the new-born human, the significance or even existence of such a mechanism in adult humans is contentious and many authorities believe that any observed NST can be attributed to pre-shivering increases in muscular tone. However, Jessen (1980) described increased metabolic rates in cooled resting adults that were accompanied by raised levels of circulating NA and FFAs. This would support a sympathetically mediated, fat lipolytic mechanism. The same author (Jessen et al., 1980) also demonstrated that increased heat production can occur in paralysed subjects after the ingestion of iced water, a situation in which increased muscular tone should be impossible and, furthermore, some studies of cold acclimatisation provide further evidence supporting the possible significance of non-shivering heat production in adult humans.

Whatever the relative significance of the behavioural and physiological responses to general body cooling, the capacity to maintain overall thermal equilibrium under harsh conditions does not necessarily imply that all parts of the body are adequately protected. Indeed, excessive cooling of the hands and feet, with consequent pain and loss of function, is generally the limiting factor in human capacity for coping with the cold.

## Physiological adaptation to cold environments

Unlike the situation regarding adaptations to heat, there are different opinions regarding humans' physiological capacity to adapt to cold. Many authorities believe that although such adaptation could theoretically include changes in body insulation and resting heat production, in reality the changes are limited to alterations in the pattern of vasoconstriction in the hands and feet. However, in some traditional societies, behavioural strategems cannot provide adequate protection and, although many studies have concluded that central cold adaptations are limited to habituation, some evidence supports the existence of physiological adaptations beyond the accepted peripheral vascular changes.

Although peripheral vasoconstriction is a valuable response to the cold, it is frequently an unwelcome response since it can cause excessive decreases in blood flow to the hands and feet, which become the limiting factors in human ability to withstand the environment. However, in many individuals partial protection against such excessive constriction is afforded by cold-induced vasodilatation (CIVD), a response whereby the constriction in cooled digital vessels is intermittently reversed (Keatinge & Harman, 1980). It is generally accepted that the capacity to defend the periphery against severe cooling can also be enhanced by repeated cold exposure and that the acclimatative change includes both a decrease in the intensity of the initial vasoconstrictor response and an increase in the frequency and duration of CIVD. Such changes have been demonstrated in Eskimos and Lapps, possibly due to evolutionary changes; however, similarly improved responses are evident in the hands of the Gaspe fisherman (LeBlanc, 1952) and fish filleters (Nelms & Soper, 1962), who are occupationally exposed to cold water.

In addition to the peripheral vascular changes, adaptations in central blood pressure responses to cold exposure have also been demonstrated in that acclimatised individuals show smaller increases in blood pressure following cold stimulation of the hands or face (LeBlanc, 1952). This could, of course, be attributed to simple habituation to the painful cooling stimuli, but this interpretation does not explain the observation that the Gaspe fishermen also demonstrated a decreased pressor response to foot cooling, despite it being documented that their feet were not occupationally exposed to cold. It would, therefore, seem that a genuine adaptation in centrally mediated vasoconstrictor responses can occur.

Both the changes in CIVD responses and the alteration in pressor responses can occur after relatively short term repeated cold exposure and both have been noted on a seasonal basis in groups such as Quebec postal workers (LeBlanc, 1975).

As noted above, several studies of traditional societies have shown that following repeated exposure to generalised cold, an individual will allow his or her skin and core temperature to drop, sometimes dramatically, without triggering normal levels of discomfort or shivering. This response has been documented in Aborigines, Bushmen, Arctic Indians and Eskimos and is generally thought to reflect their inability to protect themselves adequately through behavioural means and consequently to develop habituation to the cold (Young, 1988). Fig. 4.3 illustrates such changes in the Australian Aborigine.

Interestingly, however, similar changes have been seen on a seasonal basis in Caucasian laboratory workers (Davis & Johnston, 1961). These men were based in Kentucky, USA, using living and working accommodation that was heated in winter and cooled in summer (Fig. 4.4). Yet, they demonstrated a smaller metabolic response to a standardised cold stress

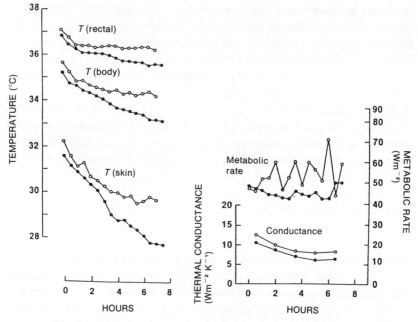

Fig. 4.3. Thermal and metabolic responses of Australian Aborigines (closed circles) and Europeans (open circles) during sleep at 5 °C with no clothing.

test during the winter than in summer, suggesting that the stimulation required to induce a degree of habituation to the cold is quite small.

Behavioural and physiological responses are relatively ineffective at maintaining thermal balance in water and many workers have used cold water to investigate responses to whole body cooling. These have generally confirmed the development of habituation with subjects acquiring a lowered threshold for the onset of shivering and a decreased intensity of shivering response at a given skin temperature (Brück *et al.*, 1976). Similar evidence has been found to occur naturally in the Ama diving women, who are regularly exposed to hypothermic stress when diving in cold water. Fig. 4.5 illustrates the variation in the degree of induced hypothermia in winter versus summer (Kang *et al.*, 1965).

Interestingly, however, not all observations of the Ama diving women can be explained by habituation alone and some results suggest that physiological adaptations in the body's insulative capacity occur (Hong, 1973). The possibility that humans are capable of such adaptations has been followed up by Young *et al.* (1986) who exposed men to 18 °C water, 5 days a week for 5 weeks. They then tested their cold responses to a 90 min. cold air stress (5 °C) arguing that while the repeated cold water immersions provided a near maximum stimulus to adaptation, a cold air stress test,

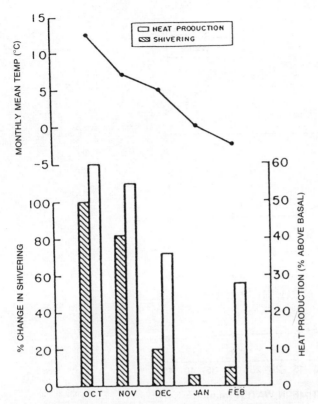

Fig. 4.4. Monthly responses to a 1 hour nude exposure to 14.1 °C air ($n=6$).

with consequent lower rates of heat loss, would provide a more discriminating assessment of possible acclimatative change. Results (Fig. 4.6) demonstrated both a hypothermic habituative response (delayed onset of shivering and similar metabolic responses despite greater cooling) and an increase in the effectiveness of body shell insulation. Since subcutaneous fat thickness did not alter, he postulated that changes in blood flow to the underlying skeletal muscle must have been responsible for this adaptive change.

Some further data suggesting an additional possible insulative adaptation have come from studies of weight change on expeditions to the Antarctic and Arctic (Stroud, 1987a, and unpublished data). The expeditions involved men, unsupported by other men, animals or machines, travelling on foot and dragging sledges containing food and equipment. In order to minimise sledge weights, the participants travelled with very limited fuel and clothing and hence generalised cold exposure occurred despite behavioural attempts to avoid it. The Antarctic journey involved

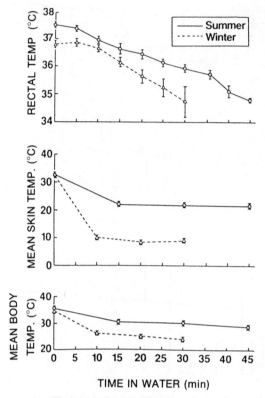

Fig. 4.5. Seasonal variation in the body temperatures of the Ama diving women (from Kang *et al.*, 1965).

three subjects walking 1400 km in 73 days (from the coast of Antarctica to the South Pole) with temperatures in the range −25 °C to +3 °C, whereas the Arctic journey involved two subjects walking 800 km in 48 days (from the northernmost point of Siberia to 88°30′ North where the attempt to reach the North Pole was abandoned due to inadequate remaining food) with temperatures in the range −45 °C to −20 °C. Despite a daily intake of 20.9 MJ in the Antarctic and 19.2 MJ in the Arctic, the men lost a lot of weight of which a disproportionate amount appeared to be lean tissues. The body composition changes are shown in Table 4.1.

The high lean weight losses in these men were very surprising and, as far as I am aware, similar findings have not been reported elsewhere. Although they may have been caused by some unrecognised effect of very sustained work loads in combination with prolonged negative energy balance, they may represent an attempt by the body to maintain subcutaneous fatty insulation.

Table 4.1. *Changes in body composition after Antarctic and Arctic expeditions*

| Expedition and subject no. | Wt loss (kg) | FFM[a] loss (kg) | Fat loss (kg) |
|---|---|---|---|
| Antarctic 1 | 11.5 | 8.6 | 2.9 |
| Antarctic 2 | 11.5 | 8.4 | 3.1 |
| Antarctic 3 | 7.9 | 6.0 | 1.9 |
| Arctic 1 | 12.8 | 5.7 | 7.1 |
| Arctic 2 | 13.5 | 7.6 | 5.7 |

[a]Fat-free mass.

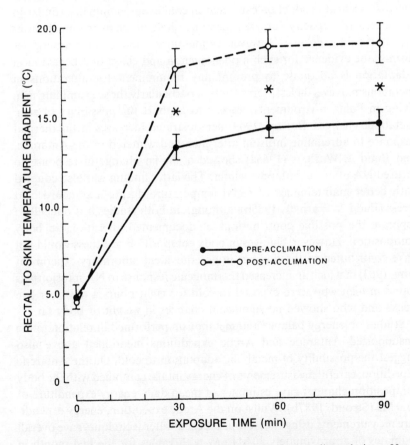

Fig. 4.6. Rectal to mean weighted skin temperature gradients during a 90 min. exposure to cold air, before and after a programme of acclimatisation using cold water ($n = 7$) * = significant difference ($p < 0.001$) (from Young *et al.*, 1988).

Table 4.2. *Changes in resting metabolic rate (RMR) following an Antarctic and Arctic expedition*

| Expedition and subject no. | RMR (Watts) | | RMR (Watts/kg FFM) | | Percentage increase in RMR (kg FMF) |
|---|---|---|---|---|---|
| | Before | After | Before | After | |
| Antarctic 1 | 87.4 | 127.0 | 1.24 | 2.04 | 64.5 |
| Antarctic 2 | 73.0 | 104.5 | 1.27 | 2.03 | 59.8 |
| Arctic 1 | 95.6 | 122.3 | 1.56 | 2.20 | 41.0 |
| Arctic 2 | 96.8 | 102.7 | 1.29 | 1.55 | 20.2 |

Various studies have attempted to establish whether significant adaptive increases in heat production can occur in cold exposed humans, similar to the increased capacity for NST demonstrated in small mammals after repeated cold exposure. However, as mentioned above, most authorities dismiss the evidence for such a phenomenon and claim that behavioural adaptation is adequate to prevent any requirement for alterations in metabolism. Nevertheless, some studies, particularly those examining men living in Polar environments, appear to support the possibility of such metabolic changes. Butson (1949) demonstrated increases in the thermal response to adrenaline infusion after men had wintered in the Antarctic, and Budd & Warhaft (1966a) showed a similar change in response to infused NA after an Antarctic winter. The latter finding was also coupled with better maintainance of rectal temperature during a standard cold stress (Budd & Warhaft, 1966b), although in both studies it is difficult to separate the possible contributions of documented alterations in body composition. However, changes in body composition or fitness would not have contributed to the findings in an equivalent laboratory experiment (Joy, 1963) in which an increased thermogenic response to NA infusion was found in men who were exposed to cold for eight hours a day over five weeks and who showed no significant changes in weight or body fat.

Studies on energy balance and metabolism performed in relation to the unsupported Antarctic and Arctic expeditions mentioned above also suggest the possibility of metabolic adaptation to cold. On the Antarctic expedition, careful measurements of energy intake combined with the body composition changes gave estimates of mean daily energy expenditure of 25.0 MJ (Stroud, 1987b). Whilst on the Arctic expedition, energy expenditure measurements using the isotope labelled water technique, gave overall estimates of approximately 30 MJ/day with values for the first month in excess of 32.0 MJ (Stroud *et al.*, in press). Such high values for sustained energy expenditure are greater than would be anticipated from studies of other hard working groups and were accompanied by elevated resting metabolic rates measured after the expeditions (Table 4.2).

Following weight loss through relative dietary restriction a decline in RMR would be anticipated, although there is some debate as to whether the fall is entirely accountable by decreases in FFM or whether it reflects a true compensatory hypometabolic response. As far as I am aware, a rise in metabolic rate following weight loss has not been reported elsewhere and although it may represent an effect of refeeding in the days following the expeditions, the high energy expenditures during the course of the expeditions themselves would suggest that the metabolic cost of both rest and work was probably raised before the refeeding stage. In addition, all measurements were made after an overnight fast and hence previous days food intake should have exerted only a small effect. It would, therefore, seem more probable that the changes were due to either an unrecognised effect of very sustained exercise or an adaptation to the cold environment.

### Conclusion

In conclusion, the physiological responses to heat and their potential for adaptation are well understood and the adaptations are highly effective in influencing human capacity to live and work in a hot environment. In view of the relatively rapid onset and decline of the adaptive changes, they are clearly evident on a seasonal basis. The physiological responses to cold, on the other hand, are less effective when it comes to climatic extremes and humans chiefly rely upon behavioural stratagems for the majority of their protection. Most authorities feel that as a consequence of the effective behavioural defences against the cold, adaptation is confined to the changes of habituation to generalised cold along with improvements in peripheral vascular responses, which permit function of the hands and feet to be maintained. There is some evidence of seasonal variation in these changes of both habituation and peripheral vascular adaptation. However, in addition to these accepted changes, there is some evidence that exposure to extreme conditions may also lead to insulative adaptations and possibly improvements in heat production.

### References

Allan, J.A. & Roddie, I.C. (1972). The role of circulating catecholamines on sweat production in man. *Journal of Physiology*, **227**, 801–14.

Bass, D.E. & Henschel, A. (1956). Responses of body fluid compartments to heat and cold. *Physiological Reviews*, **36**, 128–44.

Bazett, H.C., Sunderman, F.W., Doupe, J. & Scott, J.C. (1940). Climatic effects on the volume and composition of blood in man. *American Journal of Physiology*, **129**, 69–83.

Brück, K., Baum, E. & Schwennicki, H.P. (1976). Cold-adaptive modifications in man induced by repeated short term cold exposures and during a 10-day and -night cold exposure. *Pflügers Archiv*, **363**, 125–33.

Budd, G.M. & Warhaft, N. (1966a). Body temperature, shivering, blood pressure and heart rate during a standardised cold stress in Australia and Antarctica. *Journal of Physiology*, **186**, 216–32.

Budd, G.M. & Warhaft, N. (1966b). Cardiovascular and metabolic responses to noradrenaline in man before and after acclimatisation to cold in Antarctica. *Journal of Physiology*, **186**, 233–42.

Butson, A.R.C. (1949). Acclimatization to cold in the Antarctic. *Nature*, **163**, 132–3.

Davis, T.R.A. & Johnston, D.R. (1961). Seasonal acclimatization to cold in man. *Journal of Applied Physiology*, **16**(2), 231–4.

Hemingway, A. (1963). Shivering. *Physiological Reviews*, **41**, 397–422.

Hong, S.K. (1973). Pattern of cold adaptation in women divers of Korea (Ama). *Federation Proceedings, American Societies for Experimental Biology*, **32**, 1614–22.

Hori, S., Inouye, A., Ihzuka, H. & Yamada, T. (1974). Study on seasonal variations in heat tolerance in young Japanese males and effects of physical training thereon. *Japanese Journal of Physiology*, **24**, 463–74.

Jessen, K. (1980). An assessment of human regulatory nonshivering thermogenesis. *Acta Anaesthesiologica Scandinavica*, **24**(2), 138–43.

Jessen, K., Rabol, A. & Winkler, K. (1980). Total body and splanchnic thermogenesis in curarised man during a short exposure to cold. *Acta Anaesthiologica Scandinavica* **24**(4), 339–44.

Joy, R. (1963). Responses of cold acclimatised men to infused norepinephrine. *Journal of Applied Physiology*, **18**(6), 1209–12.

Kang, D.H., Kim, P.K., Kang, B.S., Song, S.H. & Hong, S.K. (1965). Energy metabolism and body temperature of the Ama. *Journal of Applied Physiology*, **20**, 46–50.

Keatinge, W.R. & Harman, C. (1980). Local mechanisms controlling blood vessels. *Monographs of the Physiological Society*, **37**, Chapter 8. Academic Press.

Kuhlemeier, K.V., Miller, J.M., Dukes-Dobos, F.N. & Jensen, R. (1977). Determinants of the prescriptive zone in industrial workers. *Journal of Applied Physiology*, **43**, 347–51.

Kuno, Y. (1956). The acclimatisation of the human sweat apparatus to heat. In *Human Perspiration*, ed. Y. Kuno, pp. 318–22. Springfield, IL: C.C. Thomas.

LeBlanc, J. (1952). Local adaptation to cold of Gaspe fishermen. *Journal of Applied Physiology*, **17**, 950–2.

LeBlanc, J. (1975). *Man in the Cold*. Springfield, IL: C.C. Thomas.

Lind, A.R. (1963). A physiological criterion for setting thermal environmental limits for everyday work. *Journal of Applied Physiology*, **18**, 51–6.

Martineau, L. & Jacobs, I. (1989). Muscle glycogen availability and temperature regulation in humans. *Journal of Applied Physiology*, **66**, 72–8.

Nelms, J.D. & Soper, D.J.G. (1962). Cold vasodilation and cold acclimatization in the hands of British fish filleters. *Journal of Applied Physiology*, **12**, 444–8.

Norman, J.N. (1965). Cold exposure and patterns of activity at a polar station. *British Antarctic Survey Bulletin*, **6**, 1–13.

Robinson, S. & Robinson, A.H. (1954). The chemical composition of sweat. *Physiological Reviews*, **34**, 202–20.

Sawka, N.M. & Wenger, C.B. (1988). Physiological responses to acute exercise heat stress. In *Human Performance Physiology and Environmental Medicine at Terrestrial Extremes*, ed. K.B. Pandolf, M.N. Sawka & R.R. Gonzalez, pp. 97–151. USA: Benchmark Press.

Senay, L.C. (1979). Effects of exercise in the heat on body fluid distribution. *Medicine and Science in Sports and Exercise*, **11**, 42–8.

Senay, L.C., Mitchell, D. & Wyndham, C.H. (1976). Acclimatisation in a hot, humid environment: body adjustments. *Journal of Applied Physiology*, **40**, 786–96.

Shvartz, E.H. & Meyerstein, N. (1970). Effects of heat and natural acclimatization to heat on tilt tolerance of men and women. *Journal of Applied Physiology*, **28**, 428–32.

Stroud, M.A. (1987a). Nutrition and energy balance on the 'Footsteps of Scott' expedition 1984–86. *Human Nutrition, Applied Nutrition*, **41A**, 426–33.

Stroud, M.A. (1987b). Increased basal metabolic rate after sustained exercise in a cold environment. *Lancet*, **1987**(1), 1327.

Stroud, M.A., Coward, W.A. & Sawyer, M.B. (in press). Measurements of energy expenditure using isotope labelled water ($^2H_2{}^{18}O$) during an Arctic expedition.

Tikuisis, P., Bell, D.G. & Jacobs, I. (1991). Shivering onset, metabolic response, and convective heat transfer during cold air exposure. *Journal of Applied Physiology*, **70**, 1996–2002.

Wenger, C.B. (1988). Human heat acclimatization. In *Human Performance: Physiology and Environmental Medicine at Terrestrial Extremes*, ed. K.B. Pandolf, M.N. Sawka & R.R. Gonzalez, pp. 153–97. USA: Benchmark Press.

Wyndham, C.H., Strydom, N.B., Morrison, J.F., Williams, C.G., Bredell, G.A.G., Von Rahden, M.J.E., Holdsworth, L.D., Van Graan, C.H., Van Resnberg, A.J. & Munro, A. (1964). Heat reactions of Caucasians and Bantu in South Africa. *Journal of Applied Physiology*, **19**, 598–606.

Young, A.J. (1988). Human adaptation to the cold. In *Human Performance: Physiology and Environmental Medicine at Terrestrial Extremes*, ed. K.B. Pandolf, M.N. Sawka, R.R. Gonzalez, pp. 401–34. USA: Benchmark Press.

Young, A.J., Muza, S.R., Sawka, M.N., Gonzalez, R.R. & Pandolph, K.B. (1986). Human thermoregulatory responses to cold air are altered by repeated cold water immersion. *Journal of Applied Physiology*, **60**, 1542–8.

Young, A.J., Sawka, M.N., Neufer, P.D., Muza, S.R., Askew, E.W. & Pandolph, K.B. (1989). Thermoregulation during cold water immersion is unimpaired by low muscle glycogen levels. *Journal of Applied Physiology*, **66**(4), 1890–16.

# 5 Physiological responses to variations in daylength

D.L. INGRAM AND M.J. DAUNCEY

## Introduction

Humans are subject to numerous rhythms during the year (Hildebrandt, 1966; Aschoff, 1981a, b), which impinge to various degrees on their physiology. Some of these are related to environmental temperature or nutrition and are dealt with elsewhere; the present review is concerned with the effects of daylength solely as illumination that acts through the eye on the central nervous system. For various reasons, most of the experimental evidence about the mechanisms involved in the control of rhythms by light has been obtained from studies on animals other than humans. In the following account some of this evidence will be presented and the extent to which it applies to humans will be considered.

## Annual and daily rhythms

There are at least two kinds of rhythm that can be influenced by lighting: one that occurs over a day, such as motor activity, and the other over a year, such as breeding. Both depend on structures in the brain and there is reason to believe that yearly rhythms depend at least in part on daily rhythms. The daily rhythms (i.e. nychthemeral rhythms of precisely 24 h) persist as circadian rhythms (i.e. with a period of about 24 h) even in the absence of clues about the passage of time, and a change in the timing of events (Zeitgeber) will reset the rhythm, as happens when humans move from one time zone to another.

Subjecting animals that are seasonal breeders to progressively increasing daylengths in an otherwise constant environment will either induce, or suppress, the breeding season depending on the species. Thus, ferrets can be made to begin breeding earlier in the year than is usual by exposure to lengthening days, while in sheep it is shortening days that stimulate breeding. In yet other species shortening days terminates breeding. It should, however, be pointed out that other factors such as food supply will also influence the onset of oestrus, and that breeding will eventually begin

54

even if the daylength is kept constant. Thus, although lighting can influence both circadian and circannual rhythms they also occur spontaneously. The unexpected finding with respect to breeding rhythms is that the period of illumination to which the animal is exposed does not have to be continuous; two short periods of light 11 h apart have the same effect as 11 h continuous light (Hastings *et al.*, 1985). Explanations for this phenomenon are based on external and internal coincidence models, both of which involve the circadian rhythm in the control of the annual rhythm or photoperiodic time measurement (Pittendrigh, 1972, 1981).

According to the external coincidence model, the onset of the light period entrains a circadian rhythm that marks this point as dawn. At some set interval after dawn there then occurs a light sensitive period such that if the animal experiences light during this 'window', either as a continuation of the initial illumination or as a separate burst of light, then a long day is recorded.

In the internal coincidence model it is postulated that there are at least two oscillating systems, one fixed by dawn, and the other by dusk. The phase relationship between the two oscillators could then be used to measure daylength and so detect the direction change. Both models have been used to account for different features of photoperiodic time measurement in various animals (Gwinner, 1981) but neither explains all the patterns recorded.

### Neural basis of rhythm generators

In mammals the light signals enter the nervous system only through the eye, since orbital enucleation or section of the optic nerves prevents any response of the circannual rhythm to changes in lighting (Herbert *et al.*, 1978). In both diurnal and nocturnal animals destruction of the suprachiasmatic nucleus (SCN) in the brain eliminates all the rhythms which are associated with dark/light cycles, and electrical stimulation of the SCN causes phase shifts in these rhythms that are similar to those that follow changes in the pattern of lighting (Rusak & Zucker, 1979; Rusak & Groos, 1982).

Recordings from the SCN have shown that even after its isolation from the brain there is a circadian rhythm in neural activity. Moreover, experiments in rats (nocturnal) and primates (diurnal) involving 2-deoxyglucose have demonstrated lower SCN activity in the dark than in the light. Activity within the SCN influences the pineal gland via sympathetic nerves, although not all the pathways have been identified. At least one travels via the paraventricular nucleus but its exact significance has yet to be elucidated (Hastings *et al.*, 1985).

### Role of pineal gland and melatonin

Removal of the pineal gland prevents all experimentally imposed effects of light on breeding, although animals may still breed seasonally. The gland produces the hormone melatonin (N-acetyl-5-methoxytryptamine) after stimulation of $\beta$-adrenergic receptors, which are themselves subject to a 24 h rhythm in concentration. The pineal also produces a number of other peptides, which may well prove to be important in humans as well as in other animals. However, only melatonin has been studied in detail. As soon as light is perceived by the retina, or the nervous input to the pineal gland is interrupted, production of melatonin ceases in both diurnal and nocturnal animals. Furthermore, injections of melatonin in the late afternoon can mimic the effects of short days. Melatonin production, the details of which have been reviewed recently by Reiter (1991), is closely linked with the production of the transmitter substance serotonin (5-hydroxytryptamine; 5HT), which is an intermediary product. The production of 5HT from tryptophan depends on the rate limiting enzyme tryptophan hydroxylase, the concentration of which displays a 24 h rhythm, increasing in the dark. The next stage in the formation of melatonin involves the enzyme 5HT-N-acetyl transferase, which is inhibited by exposure of the animal to light.

The inhibition of melatonin production by light occurs in both nocturnal and diurnal animals and in seasonal and non-seasonal breeders. Differences in the response of different species to light thus depend on events after reception of the hormonal signal, rather than on its specific physiological effect, for example as an antigonadotrophic agent. It should also be stressed that even in total darkness, and in blind animals including humans, there is still a circadian rhythm in melatonin production, which in humans tends to be longer than 24 h. By contrast, in animals exposed to continuous lighting, melatonin secretion is suppressed (Lewy & Newsome, 1983).

### Differences in light intensity

In many societies, individuals are exposed to some form of light during their entire waking hours and may not therefore experience any significant variation in effective daylength during the year, or from one latitude to another. In nocturnal animals, quite low levels of illumination can depress melatonin secretion and in the rat normal artificial light causes complete suppression (Reiter, 1985). In humans, by contrast, Lewy et al. (1980) found that for complete suppression of melatonin production a level of illumination of 3000 lux is needed, which is equivalent to bright sunlight, while 500 lux, which is in the range of normal artificial light, had very little effect. These findings were subsequently confirmed by McIntyre *et al.* (1989), who used exposures of one hour periods of illumination of between 200 and 300 lux during the night. As in other animals, the most effective wavelength for suppression of melatonin in humans is 500 nm.

### Pattern of melatonin secretion

The pattern of plasma melatonin concentration within an individual is fairly constant, although there are considerable variations between individuals (Broadway *et al.*, 1988; Waldhauser *et al.*, 1984). In humans, as in most species, the concentration of melatonin in plasma rises gradually soon after dark and reaches a peak in the middle of the dark period, falling again just before dawn. The episodic nature of the secretion of melatonin, which resembles that of hypothalamic releasing hormones, has also been established in humans and the half-life of the hormone found to be in the range 10–40 min. (Reiter, 1991). Shifts in the time of dusk or dawn for human volunteers changes the timing of melatonin secretion so that the hormone is always secreted in the dark phase (Lewy *et al.*, 1985).

The results of studies of melatonin concentrations in humans over the period of a year reveal systematic variations, although these are not quite in keeping with expectation. Thus Arendt *et al.* (1979) took blood samples at monthly intervals at midnight and at 08.00 hours and found two peaks, one in winter and one in summer. Confirmation of these two peaks has been reported for circulating melatonin and from studies on pineal glands removed at autopsy (Vaughan, 1984). By contrast, Wetterberg *et al.* (1981) found no variation in melatonin concentration in urine over the year. The evidence cited above, that the pattern of secretion remains fairly constant in individuals (Broadway *et al.*, 1988), also failed to suggest any differences in melatonin secretion over a long period of the year. In this study, samples were taken from subjects in the Antarctic at intervals over 24 h in winter and in the autumn and the curves of urinary excretion against time fitted almost exactly over each other. One possibility that needs further investigation under closely controlled conditions is that the effect of light interacts with environmental temperature (Vivien-Roels & Pévet, 1983), or some other variable such as physical activity or nutrient intake.

If melatonin is the hormone that mediates the effects of light on reproduction then there are at least three possible modes of action. The production of melatonin may need to coincide with a sensitive 'window', or it may be related to the time over which it is produced, or it may depend on the amplitude of production (Reiter, 1991). A detailed consideration of these possibilities is outside the scope of this review. The pathway by which melatonin acts is not yet known, although in a number of species it has been

shown that implantation of small amounts of melatonin into the hypo-
thalamus influences reproductive cycles. Most probably the effect depends
on $\beta$-endorphins, which are known to inhibit the release of GnRH. Little is
known about the influence of melatonin in humans, although it is recorded
that women with hypothalamic amenorrhoea do have high circulating
levels of the hormone (Berga et al., 1988).

### Seasonal breeding in humans

Although human populations cannot be classified as seasonal breeders like
sheep or ferrets, there are nevertheless seasonal variations in conception
rates (Aschoff, 1981b; Cowgill, 1966). The records of births in North
America, Europe and Japan (Cowgill, 1966; James, 1990) reveal a distinct
peak of conceptions in late spring and early summer; moreover, in North
America the magnitude of the increase in conceptions is correlated with the
latitude (James, 1990). Conversely, in Australia, New Zealand and South
America the peak conception rate is in December and January (Cowgill,
1966). These findings thus clearly fail to disprove the hypothesis that
human conception is influenced by the lengthening day. Close inspection of
the records, however, reveals that in individual countries the peak time for
conception tends to drift over the years in both the Southern and Northern
hemispheres (Cowgill, 1966; James, 1990; Mathers & Harris, 1983), which
would not be expected if the peak depended on daylength alone. In North
America, Europe and Australia there is also an increase in conception rates
near to Christmas, which is a time of short days in one hemisphere and long
days in the other. Furthermore, even in Nigeria where the variation in
daylength is minimal there is nevertheless a small increase in conception
rates around August (Ayeni, 1986). The correlation of conception time
with daylength thus appears to be rather weak, and if it has any
physiological significance it is not a dominant one. Moreover, there are
clearly factors other than light that vary over the year and that may
influence conception.

Another approach to the question of the effect of daylength on human
reproduction is to search for variations in the serum concentration of
reproductive hormones. In women this approach is complicated by the
monthly variations that occur, and also by the fact that high levels of
melatonin have been found in ovarian follicles (Brzezinski et al., 1987).
Since in true seasonal breeders both males and females are equally affected,
studies have therefore been concentrated on men. Smals et al. (1976)
reported peak testosterone levels in July and October with a nadir in winter
and spring; and similar results were obtained by Reinberg et al. (1978).

Kauppila *et al.* (1987) detected a decrease in testicular hormones in winter, but no peak in either the summer or late autumn, while Dabbs (1990) found a seasonal peak between November and December depending on the age of the subjects. By contrast, Huhtaniemi *et al.* (1982) found no variation of serum testosterone in 24 men living in a region near the North Pole where daylength varied between 3.5 and 22 h. Moreover, the injection of melatonin into humans has very little effect on plasma concentrations of reproductive hormones (Sizonenko & Lang, 1988). It must therefore be concluded that there is no firm evidence for a direct major role of daylength in the control of human reproduction.

### Onset of puberty

The age of puberty has been falling steadily over the last century in Northern European countries. This has coincided with the introduction of better domestic lighting and it was at one time suggested that the two events were causally connected. It has also been shown that in humans the nocturnal values of plasma melatonin decline at about the time of puberty (Waldhauser *et al.*, 1984). However, since it is now known that normal levels of domestic lighting are below those required to influence human rhythms the idea is now less attractive. Puberty in fact correlates very well with body weight and thus nutrition appears to be by far the greatest influence. On the other hand, it has been shown that blind girls reach puberty sooner than those who are normally sighted (Zacharias & Wurtman, 1964). It thus appears that lighting may well have some influence on the onset of puberty but that its effect is opposite to that which was originally suspected, although care must be taken to control for body weight when comparing results from sighted and blind girls. The mechanisms that are involved and the significance of the decline in melatonin concentration before puberty are not understood.

### Daylength changes caused by changes of time zone

Movement from one time zone to another does two things in relation to lighting: first, it presents a single day in which the duration of daylight is prolonged or shortened; second, it alters the onset and offset of the light period in relation to the body's internal clock without necessarily changing the duration of light. The effect of the first factor may not be very great: the second factor does not strictly fall within the scope of this review since the daylength is not always changed, but the effect is of considerable interest. Alterations in light/dark cycle which involve light intensities of less than

1500 lux are not effective in entrainment of human subjects unless the changes are linked with some behavioural change such as going to bed. In such studies the light change is likely to play only a minor role, since the behavioural change alone is quite effective (Wever, 1985). However, experiments in which bright light of 4000–5000 lux was used (Wever *et al.*, 1983) showed that not only were these intensities alone effective as a Zeitgeber, but that the period over which entrainment could be achieved represented 'days' as short as 18 h 20 min., or as long as 29 h 20 min., which is a much wider band than is possible using any other Zeitgeber. Moreover, there is evidence that the time taken to entrain the rhythms to a step change of 6 h is significantly reduced when bright light is used.

The potential for using bright light to reset rhythms is considerable and has been used in the treatment of free running rhythms (Hobson *et al.*, 1989) and jet lag. Treatment with melatonin has also been reported to have some effect in alleviating the effects of jet lag (Arendt *et al.*, 1987), which suggests that the effects of light are indeed mediated through the pineal gland, although these effects could be related simply to the effect of melatonin as a sleeping drug.

### Seasonal affective disorder

Variations in daylength are correlated with the occurrence of mood changes, depression and sleep disorders in some individuals; a state appropriately termed SAD (seasonal affective disorder). Patients exhibit the increase in melancholia towards the end of the winter and in early spring (Aschoff, 1981b; Rosenthal *et al.*, 1983). There is also a seasonal increase in the incidence of suicide in the late spring and early summer, which is seen in both the Northern and Southern hemispheres at the corresponding times of the year (Aschoff, 1981b). In a series of studies, patients with SAD have been subjected to either additional bright light (2500 lux) or additional dim light during the winter (Rosenthal *et al.*, 1984; Rosenthal *et al.*, 1985a, b). It was found that those who experienced a day extended by the bright light had a reduction in SAD symptoms. The administration of melatonin to patients who were still receiving extra bright light reversed some, but not all, of the beneficial effects. These studies clearly have some methodological defects, particularly because of the short half-life of melatonin and the fact that under physiological conditions it is secreted in a pulsatile fashion, but they are nevertheless of great interest.

Lewy *et al.* (1990) have explored the possibility that SAD is due to a general disruption of circadian rhythms in relation to sleep, rather than a direct effect of melatonin. The hypothesis is that most of the circadian rhythms are in phase with, and possibly governed to some extent by, levels of melatonin, but that sleep is independent of these factors. They have successfully used values of plasma melatonin to determine whether a given patient is phase advanced (has an early onset of melatonin) or phase delayed, and provided extra bright light in the evening, or morning respectively.

Not all depression is subject to seasonal variation, although some degree of desynchronization of circadian rhythms from the normal 24 h day does seem to be involved. Kripke (1985) pointed out that, in general, depressed patients have low levels of circulating melatonin, which is contrary to what might be expected from the finding that melatonin suppression by bright light may have a slight tendency to relieve the symptoms. However, although there is as yet no clear explanation for these results, it should perhaps be remembered that melatonin is not the only peptide secreted by the pineal gland and that some of the effects of light may be mediated by other peptides such as 5HT.

### Melatonin, tumours and the immune system

Melatonin has also been investigated as an oncostatic agent. Tumours are reported to spread more rapidly following the removal of the pineal gland but in rats photoperiod had no constant effect on tumour growth (Blask, 1984). However, Bartsch *et al.* (1990) have suggested that melatonin is implicated in the function of the immune system and Maestroni *et al.* (1989) have produced evidence to show that injections of melatonin into animals can increase the production of killer cells. This is an area in which more work needs to be done.

### Conclusions

In human subjects there is every reason to believe that light signals received by the eye have similar effects on the pineal gland as they do in those animals that have obvious seasonal rhythms. The differences between humans and other species are, first, that the intensity of light that is needed to reduce the secretion of melatonin or reset a circadian rhythm is much greater in humans than in other animals. Second, with respect to reproduction, the signals appear to have very little if any effect, probably because of differences in melatonin receptors. The capacity of bright light to entrain physiological circadian rhythms may, however, be of great importance in treating disorders associated with desynchronization of internal oscillators. The possible effects of light on the immune system and as an oncostatic agent are also of potential importance.

62     *D.L. Ingram and M.J. Dauncey*

## References

Arendt, J., Wirz-Justice, A., Bradtke, J. & Kornemark, M. (1979). Long-term studies on immunoreactive melatonin. *Annals of Clinical Biochemistry*, **16**, 307–12.

Arendt, J., Aldhous, M., Marks, M., Folkard, S., English, J., Marks, V. & Arendt, J.H. (1987). Some effects of jet-lag and its treatment by melatonin. *Ergonomics*, **30**, 1379–93.

Aschoff, J. (1981a). The Annual Colston Lecture. In *Biological Clocks in Seasonal Reproductive Cycles*, ed. B.K. Follet & D.E. Follet, pp. 277–88. Bristol, UK: J. Wright.

Aschoff, J. (1981b). Annual rhythms in man. In *Handbook of Behavioral Neurobiology*, Vol. 4, ed. J. Aschoff, pp. 475–87. New York: Plenum Press.

Ayeni, O. (1986). Seasonal variation of births in rural Southwestern Nigeria. *International Journal of Epidemiology*, **15**, 91–4.

Bartsch, H., Bartsch, C. & Gupta, D. (1990). Seasonal variations of endogenous defence mechanisms against cancer. In *Neuroendocrinology: New Frontiers*, ed. D. Gupta, H.A. Wollmann & M.B. Ranke, pp. 333–40. Tubingen: Brain Research Promotion.

Berga, S.L., Hortola, J.F. & Yen, S.S.C. (1988). Amplification of nocturnal melatonin secretion in women with functional hypothalamic amenorrhea. *Journal of Clinical Endocrinology and Metabolism*, **66**, 242–4.

Blask, D.E. (1984). The pineal: an oncostatic gland. In *The Pineal Gland*, ed. R.J. Relier, pp. 253–84. New York: Raven Press.

Broadway, J.W., Folkard, S. & Arendt, J. (1988). Bright light phase shifts the human melatonin rhythm in Antarctica. *Neuroscience Letters*, **79**, 185–9.

Brzezinski, A., Seibel, M.M., Lynch, J.J., Deng, M.H. & Wurtman, R.J. (1987). Melatonin in human preovulatory follicular fluid. *Journal of Clinical Endocrinology and Metabolism*, **64**, 866–7.

Cowgill, U.M. (1966). Season of birth in man, contemporary situation with special reference to Europe and the Southern hemisphere. *Ecology*, **47**, 614–23.

Dabbs, J.M. (1990). Age and seasonal variation in serum testosterone concentration among men. *Chronobiology International*, **7**, 245–9.

Gwinner, E. (1981). Circannual Systems. In *Handbook of Behavioral Neurobiology, Vol. 4, Biological Rhythms*, ed. J. Aschoff, pp. 391–410. New York: Plenum.

Hastings, H.H., Herbert, J., Martensz, N.D. & Roberts, A.C. (1985). Annual reproductive rhythms in mammals: mechanisms of light synchronization. In *The Medical & Biological Effects of Light, Annals of the New York Academy of Sciences*, Vol. **453**, ed. R.J. Wurtman, H.J. Baum & J.T. Potts, pp. 182–204. New York Academy of Sciences.

Herbert, J., Stacey, P.M. & Thorpe, D.H. (1978). Recurrent breeding seasons in pinealectomized or optic sectioned ferrets. *Journal of Endocrinology*, **78**, 389–97.

Hildebrandt, G. (1966). Biologische Rhythmen und ihre Bedeutung für Bäder-und Klimaheilkunde. In *Handbuch der Bäder und Klimaheilkunde*, ed. A. Amelung & A. Evers, pp. 730–85. Stuttgart: Schattauer Verlag.

Hobson, T.H., Sack, R.L., Lewy, A.J., Miller, L.S. & Singer, C.M. (1989). Entrainment of a free running human with bright light. *Chronobiology International*, **6**, 347–53.

Huhtaniemi, I., Martikainen, H. & Tapanainen, J. (1982). Large annual variation in photoperiodicity does not affect testicular endocrine function in man. *Acta Endocrinologica*, **101**, 105–7.

James, W.H. (1990). Seasonal variation in human births. *Journal of Biosocial Sciences*, **22**, 113–19.

Kauppila, A., Kivela, A., Pakarinen, A. & Vakkuri, D. (1987). Inverse seasonal relationship between melatonin and ovarian activity in humans in a region with a strong seasonal contrast in luminosity. *Journal of Clinical Endocrinology and Metabolism*, **65**, 823–8.

Kripke, D.F. (1985). Therapeutic effects of bright light in depressed patients. In *The Medical & Biological Effects of Light, Annals of the New York Academy of Sciences*, **453**, ed. R.J. Wurtman, H.J. Baum & J.T. Potts, pp. 270–81. New York Academy of Sciences.

Lewy, A.J. & Newsome, D.A. (1983). Different types of melatonin circadian rhythms in some blind subjects. *Journal of Clinical Endocrinology and Metabolism*, **56**, 1103–7.

Lewy, A.J., Wehr, T.A., Goodwin, F.K., Newsome, D.A. & Hankey, S.P. (1980). Light suppresses melatonin secretion in humans. *Science*, **210**, 1267–9.

Lewy, A.J., Sack, R.L. & Singer, C.M. (1985). Immediate and delayed effects of bright light on human melatonin production: Shifting 'Dawn' and 'Dusk' shifts the dim light melatonin onset (DLMO). In *Medical & Biological Effects of Light. Annals of the New York Academy of Sciences*, **453**, ed. R.J. Wurtman, Baum, H.J. & J.T. Potts, pp. 253–9. New York Academy of Sciences.

Lewy, A.J., Sack, R.L. & Singer, C.M. (1990). Bright light, melatonin and winter depression: The phase shift hypothesis. In *Biological Rhythms, Mood Disorders, Light Therapy and The Pineal Gland*, ed. M. Shafii & S.L. Shafii, pp. 141–73. Washington: American Psychiatric Press.

Maestroni, G.J.M., Conti, A. & Pierpoali, W. (1989). Melatonin, stress and the immune system. In *Pineal Research Reviews*, Vol. 7, ed. R.J. Reiter, pp. 203–26. New York: Alan R. Liss.

Mathers, C.D. & Harris, R.S. (1983). Seasonal distribution of births in Australia. *International Journal of Epidemiology*, **12**, 326–31.

McIntyre, I.M., Norman, T.R., Burrows, G.D. & Armstrong, S.M. (1989). Human melatonin suppression by light is intensity dependent. *Journal of Pineal Research*, **6**, 149–56.

Pittendrigh, C.S. (1972). Circadian surfaces in the diversity of possible roles of circadian organisation in photoperiodic induction. *Proceedings of the National Academy of Sciences*, **69**, 2734–7.

Pittendrigh, C.S. (1981). Circadian organisation and photoperiodic phenomena. In *Biological Clocks in Seasonal Reproductive Cycles*, ed. B.K. Follet & D.E. Follet, pp. 1–35. Bristol: Scientechnica.

Reinberg, A., Lagoguey, M., Cesselin, F., Touitou, Y., Legrand, J-C., Delassalle, A., Antreassian, J. & Lagoguey, A. (1978). Circadian and circannual rhythms in plasma hormones and other variables of five healthy young human males. *Acta Endocrinologica*, **88**, 417–27.

Reiter, R.J. (1985). Action spectra, dose-response relationships, and temporal aspects of lights effects on the pineal gland. In *The Medical & Biological Effects of Light. Annals of the New York Academy of Sciences*, **453**, ed. R.J. Wurtman, H.J. Baum & J.T. Potts, pp. 215–30. New York Academy of Sciences.

Reiter, R.J. (1991). Pineal melatonin: Cell biology of its synthesis and of its physiological interactions. *Endocrine Reviews*, **12**, 151–80.

Rosenthal, N.E., Sack, D.A., Gillin, J.C., Lewry, A.J., Goodwin, F.K., Davenport, Y., Mueller, P.S., Newsome, D.A. & Wehr, T.A. (1984). Seasonal affective disorder: A description of the syndrome and preliminary findings with light therapy. *Archives of General Psychiatry*, **41**, 72–80.

Rosenthal, N.E., Sack, D.A. & Wehr, T.A. (1983). Seasonal variation in affective disorders. In *Circadian rhythms in Psychiatry*, ed. T.A. Wehr & F.K. Goodwin, pp. 185–201. Pacific Grove, CA: Boxwood Press.

Rosenthal, N.E., Sack, D.A., James, S.P., Parry, B.L., Mendelson, W.B., Tamarkin, L. & Wehr, T.A. (1985a). Seasonal affective disorder and phototherapy. In *The Medical & Biological Effects of Light. Annals of the New York Academy of Sciences*, Vol. **453**, ed. R.J. Wurtman, H.J. Baum & J.T. Potts, pp. 260–9. New York Academy of Sciences.

Rosenthal, N.E., Sack, D.A., Carpenter, C.J., Parry, B.L., Mendelson, W.B. & Wehr, T.A. (1985b). Antidepressant effect of light in seasonal depression. *American Journal of Psychiatry*, **142**, 163–70.

Rusak, B. & Groos, G. (1982). Suprachiasmatic stimulation phase shifts circadian rhythms. *Science*, **215**, 1407–9.

Rusak, B. & Zucker, I. (1979). Neural regulation of circadian rhythms. *Physiological Reviews*, **59**, 449–526.

Sizonenko, P.C. & Lang, U. (1988). Melatonin and human reproduction. In *Melatonin: Clinical Perspectives*, ed. A. Miles, D.R.S. Philbrick & C. Thompson, pp. 62–78. Oxford University Press.

Smals, A.G.H., Kloppenborg, P.W.C. & Benraad, T.J. (1976). Circannual cycle in plasma testerone levels in man. *Journal of Clinical Endocrinology and Metabolism*, **42**, 979–82.

Vaughan, G.M. (1984). Melatonin in humans. In *Pineal Research Reviews*, Vol. **2**, ed. R.J. Reiter, pp. 141–201. New York: Alan R. Liss.

Vivien-Roels, B. & Pévet, P. (1983). The pineal gland and the synchronization of reproductive cycles with variations of the environmental climatic conditions, with special reference to temperature. In *Pineal Research Reviews*, Vol. **1**, ed. R.J. Reiter, pp. 91–143. New York: Alan R. Liss.

Waldhauser, F., Weissenbacher, U., Zeitlhuber, M., Waldhauser, H.F., Frisch, R. & Wurtman, R.J. (1984). Fall in nocturnal serum melatonin levels during puberty and pubescence. *Lancet*, **i**, 362–5.

Wetterberg, L., Halberg, F., Haus, E., Kawaski, T., Uezono, K., Ueno, H. & Omae, T. (1981). Circadian rhythmic melatonin in four seasons by clinically healthy Japanese subjects in Kyushu. *Chronobiologia*, **8**, 188–9.

Wever, R.A. (1985). Use of light to treat jet lag: Differential effects of normal and bright artificial light on human circadian rhythms. In *The Medical & Biological Effects of Light. Annals of the New York Academy of Sciences*, Vol. **453**, ed. R.J. Wurtman, H.J. Baum & J.T. Potts, pp. 282–304. New York Academy of Sciences.

Wever, R.A., Polasek, J. & Wildgruber, C.M. (1983). Bright light affects human circadian rhythms. *Pflügers Archiv*, **396**, 85–7.

Zacharias, L. & Wurtman, R.J. (1964). Blindness: its relation to age of menarche. *Obstetrics and Gynecology*, **30**, 507–9.

# 6 Seasonality and fertility

LYLIANE ROSETTA

## Introduction

The main seasonal factors that can influence fertility in a number of mammalian species are photoperiod, temperature and humidity. Although traces of the same kind of regulation are likely to be found in most species, human beings differ from most other species in that they are socialized and are non-seasonal breeders, and it is worth asking whether these seasonal factors influence fertility in humans.

This chapter will review recent observations about circannual variability of a number of measures of human fertility, including sperm quality, menstrual cycle cyclicity and hormonal rhythmicity, in addition to factors influencing birth seasonality. The ways in which climatic seasonality, as mediated through variations in photoperiod, temperature or humidity, can interfere with the physiological processes associated with fertility are also examined, as is the importance of interactions between environmental factors and social life in determining human fertility.

## Seasonal variability in fertility parameters

### Sperm quality

The last decade has seen an increase in medically-assisted conception in developed countries, and the number of sperm banks has risen as a consequence. Studies of sperm quality, carried out across the calendar year, show seasonal fluctuations in semen quality for both fertile and infertile males.

Analyses of semen from fertile volunteer donors in Lille (France) have shown seasonal variability in sperm count, with the highest values recorded in late winter and early spring and the lowest values recorded in late summer (Saint Pol et al., 1989). No circannual pattern was demonstrated either for sperm volume or for percentage of motile spermatozoa. A similar finding has been reported in the United States. Sperm analysis of 17 837 ejaculates donated at the California cryobank between April 1988 and

65

December 1990 showed highest sperm quality to be in December and January, with lower quality in June, July, August and September (Broder *et al.*, 1991).

### Hormonal rhythmicity

The influence of primary testicular failure on circannual hormone rhythmicity has been examined by determining the annual rhythms for the secretion of testosterone, luteinizing hormone (LH), follicle stimulating hormone (FSH) and prolactin (PRL) in healthy fertile men (Bellastella *et al.*, 1986). A peak in testosterone was detected in late September, and a small rise in LH in February and FSH in January. PRL values did not show any significant rhythmicity. These results are in agreement with other studies (Touitou *et al.*, 1983).

A study of seasonality in the birth of boys suffering from cryptorchidism (incomplete testicular descent), in the Oxfordshire Health District during the years 1974–83 (Jackson & Swerdlow, 1986) shows a peak in April, a time of low testosterone production in males. The authors suggested that a testosterone surge, at or just after the time of birth, may be necessary for normal testicular descent. Thus it may be that infants born in spring have a reduced testosterone surge.

### Menstrual pattern

Sundararaj and coworkers (1978) have analyzed the menstrual histories of 3800 women living in Minnesota in an attempt to explore seasonality of menstrual characteristics – mean menstrual interval lengths, menstrual variability (as measured by the standard deviation of menstrual interval length), and the number of menstrual cycles per woman per season. A recurrent sinusoidal pattern for cycle length was found, decreasing during the warmer days of spring and summer, and increasing in autumn. However, the authors concluded that the variations, even if statistically significant, 'were not large enough to indicate any physiological relationship'.

### Birth seasonality

Numerous studies of birth seasonality have been carried out, going back to the seventeenth century (Cowgill, 1966). Most of these have been retrospective, or historical in nature. In most countries, basic data are available for the number of births recorded by health authorities. Such data can be used to examine birth seasonality. For example, in Kenya, Ferguson

(1987) examined registered births data gathered at district level over a period of 42 months, between 1979 and 1982. All regions of the country were included apart from the relatively sparsely populated arid and semi-arid lands of the North. There was a large peak in registered births in September, with a secondary peak around April–May. The peak in September would correspond with high conception rates during December of the previous year, a time when many male urban workers rejoin their wives in the rural areas. While the second peak in April–May would correspond with conception during the main harvest period of July–August.

In the three urban districts of Nairobi, Nyeri and Mombasa, the trend was slightly different, with a consistent peak in March–May and a corresponding trough in November–January. Ferguson hypothesized that rainfall and temperature were likely to be important factors influencing this trend. Birth seasonality in Mombasa showed a strong negative correlation with the mean minimum temperatures occurring nine months before the month of birth, similar negative relationships also being found in 21 of the 30 districts examined.

In a study of male fertility conducted in Coast Province, the hottest region of Kenya, it was shown that the sperm concentration and total sperm counts of 30 normal male subjects were within normal limits (Rogo *et al.*, 1985), suggesting that there was no inhibition of spermatogenesis due to heat stress. Alternatively, the relationship between birth seasonality and rainfall could operate through food availability, although such a relationship has yet to be demonstrated.

A side effect of circannual cycling of temperature and humidity is the variation in incidence of those infectious and parasitic diseases that might interfere with fertility. This has been investigated by Miura and co-workers, first of all in Japan, and then in Canada (Miura, 1987; Nonaka *et al.*, 1990). They hypothesized that seasonality of birth could be mainly the result of seasonal infections, which cause infertility by early abortion of embryos. According to them, babies surviving the period of high prevalence of infection are likely to be the strongest and will have acquired an immunity that will last for at least 30 years and that will enable them to give birth during periods of massive infectious disease prevalence. Non-immune mothers, however, are more likely to have an early abortion. Analysis of paired data of birth-month of mothers and their respective offspring showed that births from mothers born in May–July showed very little seasonal variation while births from mothers born at other times of year had great seasonal variations, associated with higher incidence of seasonal abortions. This was observed in both the Japanese and French-Canadian populations studied.

The explanation offered by Miura and co-workers seems implausible, since it is unlikely that the levels of infection needed to cause early abortion would go unreported in most parts of the world, let alone Japan and Canada. Further, there is some evidence of hormonal dysfunction in populations with a low fertility rate that cannot be attributed to infectious factors (Wood *et al.*, 1985; Ellison *et al.*, 1986; Bailey *et al.*, 1992).

It seems more reasonable to consider a multivariate causal effect on birth seasonality with both environmental and behavioural factors playing a role. Condon & Scaglion (1982) summarized the components of birth seasonality after studying two environmentally and ethnographically distinct societies: the Copper Inuit of the Central Canadian Arctic and the Samukundi Abelam of Papua New Guinea. They concluded that 'birth seasonality may be the result of the independent action of biorhythms and sociorhythms, or, alternately, a consequence of an interaction between the two'. Studies of birth seasonality in various geographic regions highlight the importance of such climatic variables as heat stress, high humidity, and prolonged winter darkness in decreasing successful conception rates (Becker *et al.*, 1986; Condon, 1982; Ehrenkranz, 1983), although it must be acknowledged that there is an interaction between physiological and behavioural factors influencing fertility (Stoeckel & Chowdhury, 1980; Tembon, 1990).

### Climatic factors possibly influencing fertility

#### Photoperiod

It is still not clear whether the photoperiod, known to influence the reproductive function of many animal species through the pineal gland and the secretion of melatonin during darkness, also plays a role in human fertility. In seasonal breeding animals, this type of regulation plays a major part in the circannual rhythmicity of reproductive life (Ebling & Lincoln, 1987; Yeoman *et al.*, 1988; Nozaki *et al.*, 1991).

In humans, there has recently been an increase in interest in the role of melatonin (Brzezinski *et al.*, 1988; Strassman *et al.*, 1989). In particular, researchers are interested in determining whether there are specific melatonin receptors in the ovary, and if there is a stimulating effect of melatonin on the secretion of ovarian progesterone. According to Rönnberg *et al.* (1990), melatonin occurs in higher concentrations in human preovulatory follicular fluid than in serum, both in stimulated and spontaneous cycles, with significant circadian and circannual variations, suggesting a possible role for melatonin in the regulation of the human reproductive function at the follicular level. This is supported by the finding that during the arctic winter, the duration of the nocturnal melatonin secretion is increased in women and accompanied by impaired ovarian function (Kauppila *et al.*, 1987).

Male volunteers taking part in a study in which the duration of daylength was moved from summer-type to winter-type photoperiodicity showed a change in sleep pattern, becoming bimodal during the long nights with a tendency to sleep longer. In six out of seven subjects who completed the experiment, the duration of nocturnal melatonin secretion lengthened when the photoperiod was shortened (Wehr, 1991). This suggests that humans still having the ability to modify their melatonin secretion in response to changes in photoperiod may show associated changes in physiology and behaviour linked to melatonin secretion.

An elevation of daytime melatonin level has been found in highly trained sportswomen, regardless of whether they were amenorrhoeic or not (Laughlin *et al.*, 1991). Further, those with amenorrhoea had augmented nocturnal peaks of melatonin secretion with values twice those of women with normal menstrual cycles.

The mechanism by which hypersecretion of melatonin in athletic women takes place is still unknown. Laughlin *et al.* (1991) failed to show any modification in the melatonin secretion after using opioidergic and dopaminergic blockade with naloxone and metaclopramine, and it remains to be seen if there is an hypothalamic role of melatonin in the regulation of gonadotrophin releasing hormone (GnRH) pulsatile secretion. In another experiment, the same group observed an increased LH pulse amplitude in response to exogenous melatonin administration during the early follicular phase of the menstrual cycle, whereas LH pulse frequency, serum FSH and ovarian steroids were not altered (Cagnacci *et al.*, 1991).

While melatonin is secreted in the absence of visible light, vitamin D is secreted after exposure to the short-wave components of sunlight. Vitamin D (1,25-dihydroxycholecalciferol) is a steroid hormone (also called soltriol), which may be involved in the modulation of LH secretion. Premature female infants undergoing phototherapy for jaundice have been shown to have high levels of FSH and LH at 3 weeks of age, which decrease three weeks later (Lemaitre *et al.*, 1979). Another group of infants receiving the same therapy with their eyes covered showed a delay in the increase in LH levels (Dacou-Voutetakis *et al.*, 1978), suggesting that two different pathways for vitamin D activity exist: one due to short-wave radiation exposure to the skin, and another due to exposure to the eyes. In humans, there is a rise in circulating soltriol concentration during puberty, which appears to be paralleled by a decrease in melatonin levels (Aksnes & Aarskog, 1982).

Results from autoradiographic studies with [3H] soltriol have revealed many target organs and cell types with nuclear receptors for soltriol; these include female and male reproductive organs, pituitary, brain, adrenal medulla, $\beta$-cells in the endocrine pancreas, and skin (Stumpf & Denny, 1989).

### Temperature

A direct role of heat on the quality of sperm has long been suspected (Kandeel & Swerdloff, 1988) while an indirect effect of ambient temperature on fertility may operate through variability in coital frequency (Guptill *et al.*, 1990).

The role of heat on semen quality was examined in outdoor workers in San Antonio, Texas during summer and winter, when the average maximal daily temperature was 35.9 °C in July and August 1986 and 17.7 °C in January and February 1987 (Levine *et al.*, 1990). During the summer of 1986, the subjects spent an average of eight hours per day working outdoors or in settings that were not air-conditioned. The sperm concentration, total sperm count per ejaculate, and motile-sperm concentration were all significantly lower in summer than in winter. The proportion of men with sperm concentration below 20 million per millilitre (oligospermia) increased from 1 out of 131 in winter to 13 out of 131 in summer. A clear deficit in the mean number of births has been observed during spring among couples where wives lived within 250 miles of San Antonio and it is possible that this is related to the lower sperm quality of fathers during the corresponding summer. The causal effect of temperature on sperm quality has since been debated, some workers arguing that a similar phenomenon observed in the cooler climates of Lille or Basel could be related to photoperiod rather than temperature (Snyder, 1990).

### Humidity

If there is an effect of humidity on human reproduction, it is likely to be an indirect one. It is likely to operate through seasonal variability in rainfall, which is linked to social activities and socio-economic factors known to affect fecundability, such as food availability, workload, migration and vacations.

Most studies of birth seasonality carried out in equatorial or tropical areas have shown an indirect role of humidity on fertility. This operates through agricultural seasonality, which is dependent on rainfall, with its consequences for food availability, physical workload and exposure to infectious or parasitic disease, malaria in particular (Sindiga, 1987; Bantje, 1988; Leslie & Fry, 1989).

Another aspect of social life linked to rainfall is labour migration, which involves male workers from poor rural areas coming back into the village to participate in agricultural work during the rainy season; during the rest of the year opportunities to come back for brief periods are few (Huss-Ashmore, 1988).

### Discussion

Even if the mechanisms and the mediators involved in the seasonal regulation of fertility in human beings are still unknown, there is evidence of physiological seasonal variability operating through variation in sperm quality in men, and in mean duration of the menstrual cycle in women. This is true for well-nourished populations, as well as under-nourished ones. However, this may be biological synchronization caused by environmental cues rather than a real fall in fertility potential.

The trend towards an increasingly artificial lifestyle in terms of temperature or light exposure can only reduce the influence of environmental factors on fertility (Kallan & Udry, 1989; Arcury *et al.*, 1990; James, 1990). Those populations that are most exposed to external conditions are those with a 'natural' way of life, usually those experiencing poor living conditions, and directly dependent on the land and the climate for their food resources. Under such conditions the factors influencing fertility are less likely to be rainfall and temperature than nutrition and physical workload (Mosher, 1979).

A combination of seasonal food shortages and high levels of physical activity, if acute, may well regulate reproductive function in women. There is evidence to suggest that the quality and quantity of food intake may play a role in such regulation. At a peripheral level, food deficiencies and a vegetarian diet may act to impair oestradiol metabolism, mainly by increasing the faecal excretion of active metabolites (Goldin *et al.*, 1986). Simultaneously, such a diet seems to favour an alternative metabolic pathway, producing catecholestrogens, which are inactive analogues of oestradiol and competitors to it, at the pituitary level (Longcope *et al.*, 1987). At the central level, a recent study carried out on male Rhesus monkeys has shown that even short term food restriction is able to slow down, and if maintained possibly stop, the frequency of pulsatile GnRH secretion (Cameron & Nosbisch, 1991). The same phenomenon has been observed in a study of healthy men, who after 48 hours of fasting showed a serious reduction in the release of LH and consequent testosterone secretion (Cameron *et al.*, 1991).

Many rural populations living in tropical or equatorial areas have heavy physical workloads during the rainy season. If such work is considered to

72    *Lyliane Rosetta*

be equivalent to harsh endurance exercise, then a comparison can be made with athletes, in whom daily bouts of high intensity training lasting between 30 and 40 minutes have been shown to induce the secretion of beta-endorphin at hypothalamic level, an effect that may impair the pulsatility of GnRH secretion (Rosetta, 1993).

The refore, nutritional deficiencies and high levels of physical work may operate together in impairing reproductive function, breastfeeding serving to reinforce or to prolong the period of impairment (Rosetta, 1989). This pattern is evident in contrasting climates and in poor rural populations directly dependent on natural resources for their subsistence. It should not be ignored, however, that behavioural factors, in addition to biological factors, are also important in the regulation of birth seasonality.

### References

Aksnes, L. & Aarskog, D. (1982). Plasma concentrations of vitamin D metabolites in puberty: effects of sexual maturation and implications for growth. *Journal of Clinical Endocrinology and Metabolism*, **55**(1), 94–101.

Arcury, T.A., Williams, B.J. & Kryscio, R.J. (1990). Birth seasonality in a rural U.S. county, 1911–1979. *American Journal of Human Biology*, **2**, 675–89.

Bailey, R.C., Jenike, M.R., Ellison, P.T., Bentley, G.R., Harrigan, A.H. & Peacock, N.R. (1992). The ecology of birth seasonality among agriculturalists in central Africa. *Journal of Biosocial Science*, **24**(3), 393–412.

Bantje, H.F.W. (1988). Female stress and birth seasonality in Tanzania. *Journal of Biosocial Science*, **20**, 195–202.

Becker, S., Chowdhury, A. & Leridon, H. (1986). Seasonal patterns of reproduction in Matlab, Bangladesh. *Population Studies*, **40**(3), 457–72.

Bellastella, A., Criscuolo, T., Sinisi, A.A., Iorio, S., Sinisi, A.M., Rinaldi, A. & Faggiano, M. (1986). Circannual variations of plasma testosterone, luteinizing hormone, follicle-stimulating hormone and prolactine in Klinefelter's syndrome. *Neuroendocrinology*, **42**, 153–7.

Broder, S., Rothman, C. & Sims, C. (1991). A $2\frac{1}{2}$ year, 300 donor study confirms seasonal variation in sperm quality. Abstract. In *Fertility and Sterility, 1991, Program Supplement, 47th Annual Meeting of the American Fertility Society*, p. S–103.

Brzezinski, A., Lynch, H.J., Seibel, M.M., Deng, M.H., Nader, T.M. & Wurtman, R.J. (1988). The circadian rhythm of plasma melatonin during the normal menstrual cycle and in amenorrheic women. *Journal of Clinical Endocrinology and Metabolism*, **66**(5), 891–5.

Cagnacci, A., Elliott, J.A. & Yen, S.S.C. (1991). Amplification of pulsatile LH secretion by exogenous melatonin in women. *Journal of Clinical Endocrinology and Metabolism*, **73**(1), 210–2.

Cameron, J.L. & Nosbisch, C. (1991). Suppression of pulsatile luteinizing hormone and testosterone secretion during short term food restriction in the adult male rhesus monkey (*Macaca mulatta*). *Endocrinology*, **128**, 1532–40.

Cameron, J.L., Weltin, T.E., McConaha, C., Helmreich, D.L. & Kaye, W.H. (1991). Slowing of pulsatile luteinizing hormone secretion in men after forty-eight hours of fasting. *Journal of Clinical Endocrinology and Metabolism*, **73**, 35–41.

Condon, R.G. (1982). Inuit natality rhythms in the central Canadian Arctic. *Journal of Biosocial Science*, **14**, 167–77.

Condon, R.G. & Scaglion, R. (1982). The ecology of human birth seasonality. *Human Ecology*, **10**(4), 495–511.

Cowgill, U.M. (1966). Historical study of the season of birth in the city of York, England. *Nature*, **209**, 1067–70.

Dacou-Voutetakis, C., Anagnostakis, D. & Matsaniotis, N. (1978). Effect of prolonged illumination (Phototherapy) on concentrations of luteinizing hormone in human infants. *Science*, **199**, 1229–31.

Ebling, F.J.P. & Lincoln, G.A. (1987). β-endorphin secretion in Rams related to season and photoperiod. *Endocrinology*, **120**(2), 809–18.

Ehrenkranz, J.R. (1983). Seasonal breeding in humans: birth records of the Labrador Eskimo. *Fertility and Sterility*, **40**, 485–9.

Ellison, P.T., Peacock, N.R. & Lager, C. (1986). Salivary progesterone and luteal function in two low-fertility populations of Northeast Zaire. *Human Biology*, **58**(4), 473–83.

Ferguson, A.G. (1987). Some aspects of birth seasonality in Kenya. *Social Science and Medicine*, **25**(7), 793–801.

Goldin, B.R., Adlercreutz, H., Gorbach, S.L., Woods, M.N., Dwyer, J.T., Conlon, T., Bohn, E. & Gershoff, S.N. (1986). The relationship between estrogen levels and diets of Caucasian American and Oriental immigrant women. *American Journal of Clinical Nutrition*, **44**, 945–53.

Guptill, K., Berendes, H., Forman, M.R., Chang, D., Sarov, B., Naggan, L. & Hundt, G.L. (1990). Seasonality of births among bedouin arabs residing in the Negev desert of Israel. *Journal of Biosocial Science*, **22**, 213–23.

Huss–Ashmore, R. (1988). Seasonal patterns of birth and conception in rural highland Lesotho. *Human Biology*, **60**(3), 493–506.

Jackson, M.B. & Swerdlow, A.J. (1986). Seasonal variations in cryptorchidism. *Journal of Epidemiology and Community Health*, **40**, 210–13.

James, W.H. (1990). Seasonal variation in human births. *Journal of Biosocial Science*, **22**, 113–19.

Kallan, J.E. & Udry, J.R. (1989). Demographic components of seasonality of pregnancy. *Journal of Biosocial Science*, **21**, 101–8.

Kandeel, F.R. & Swerdloff, R.S. (1988). Role of temperature in regulation of spermatogenesis and the use of heating as a method for contraception. *Fertility and Sterility*, **49**, 1–23.

Kauppila, A., Kivelä, A., Pakarinen, A. & Vakkuri, O. (1987). Inverse seasonal relationship between melatonin and ovarian activity in humans in a region with a strong seasonal contrast in luminosity. *Journal of Clinical Endocrinology and Metabolism*, **65**, 823–8.

Laughlin, G.A., Loucks, A.B. & Yen, S.S.C. (1991). Marked augmentation of nocturnal melatonin secretion in amenorrheic athletes, but not in cycling athletes: unaltered by opioidergic or dopaminergic blockade. *Journal of Clinical Endocrinology and Metabolism*, **73**(6), 1321–6.

Lemaitre, B.J., Toubas, P.L., Dreux, C. & Minkowski, A. (1979). Increased gonadotropin levels in newborn premature females treated by phototherapy. *Journal of Steroid Biochemistry*, **10**, 335.

Leslie, P.W. & Fry, P.H. (1989). Extreme seasonality of births among nomadic Turkana pastoralists. *American Journal of Physical Anthropology*, **79**, 103–15.

Levine, R.L., Mathew, R.M., Brandon Chenault, C., Brown, M.H., Hurtt, M.E., Bentley, K.S., Mohr, K.L. & Working, P.K. (1990). Differences in the quality of semen in outdoor workers during summer and winter. *The New England Journal of Medicine*, **323**, 12–16.

Longcope, C., Gorbach, S., Goldin, B., Woods, M., Dwyer, J., Morrill, A. & Warram, J. (1987). The effect of a low fat diet on estrogen metabolism. *Journal of Clinical Endocrinology and Metabolism*, **64**(6), 1246–50.

Miura, T. (1987). Causes and effects of birth seasonality: general considerations. *Progress in Biometeorology*, **6**, 1–12.

Mosher, S.W. (1979). Birth seasonality among peasant cultivators: the interrelationship of workload, diet, and fertility. *Human Ecology*, **7**(2), 151–81.

Nonaka, K., Desjardins, B., Légaré, J., Charbonneau, H. & Miura, T. (1990). Effects of maternal birth season on birth seasonality in the Canadian population during the Seventeenth and Eighteenth centuries. *Human Biology*, **62**(5), 701–17.

Nozaki, M., Watanabe, G. & Taya, K. (1991). Marked seasonal changes in response to the negative feedback action of estradiol on luteinizing hormone secretion in the female Japanese Monkey. *Endocrinology*, **128**(3), 1291–7.

Rogo, K.A., Sekadde-Kigondu, C.B., Muitta, M.N. *et al.* (1985). The effects of tropical conditions on male fertility indices. *Journal of Obstetrics and Gynaecology of Eastern and Central Africa*, **4**, 45.

Rönnberg, L., Kauppila, A., Leppäluoto, J., Martikainen, H. & Vakkuri, O. (1990). Circadian and seasonal variation in human preovulatory follicular fluid melatonin concentration. *Journal of Clinical Endocrinology and Metabolism*, **71**(2), 493–6.

Rosetta, L. (1989). Breast-feeding and post-partum amenorrhea in Serere women in Senegal. *Annals of Human Biology*, **16**, 311–20.

Rosetta, L. (1993). Female reproductive dysfunction and intense physical training. *Oxford Reviews of Reproductive Biology*, **15**, 113–41.

Saint Pol, P., Hermand, E., Beuscart, R., Jablonski, W. & Leroy-Martin, B. (1989). Circannual rhythms of sperm parameters of fertile men. *Fertility and Sterility*, **51**(6), 1030–3.

Sindiga, I. (1987). Fertility control and population growth among the Maasai. *Human Ecology*, **15**(1), 53–66.

Snyder, P.J. (1990). Fewer sperm in the summer. It's not the heat, it's . . . [Editorial, comment] *The New England Journal of Medicine*, **323**, 54–6.

Stoeckel, J. & Chowdhury, A.K.M.A. (1980). Fertility and socio-economic status in rural Bangladesh: differentials and linkages. *Population Studies*, **34**(3), 519–24.

Strassman, R.J., Appenzeller, O., Lewy, A.J., Qualls, C.R. & Peake, G.T. (1989). Increase in plasma melatonin, $\beta$-endorphin, and cortisol after a 28.5-mile mountain race: relationship to performance and lack of effect of naltrexone. *Journal of Clinical Endocrinology and Metabolism*, **69**(3), 540–5.

Stumpf, W.E. & Denny, M.E. (1989). Vitamin D (soltriol), light, and reproduction. *American Journal of Obstetrics and Gynecology*, **161**, 1375–84.

Sundararaj, N., Chern, M., Gatewood, L., Hickman, L. & McHugh, R. (1978). Seasonal behavior of menstrual cycles: a biometric investigation. *Human Biology*, **50**(1), 15–31.

Tembon, A.C. (1990). Seasonality of births in the North West Province, Cameroon: implications for family planning programme. *Central African Journal of Medicine*, **36**(4), 90–3.

Touitou, Y., Lagoguey, M., Bogdan, A., Reinberg, A. & Beck, H. (1983). Seasonal rhythms of plasma gonadotrophins: their persistence in elderly men and women. *Journal of Endocrinology*, **96**, 15–23.

Wehr, T.A. (1991). The durations of human melatonin secretion and sleep respond to changes in daylength (photoperiod). *Journal of Clinical Endocrinology and Metabolism*, **73**(6), 1276–80.

Wood, J.W., Johnson, P.L. & Campbell, K.L. (1985). Demographic and endocrinological aspects of low natural fertility in highland New Guinea. *Journal of Biosocial Science*, **17**, 57–79.

Yeoman, R.R., Aksel, S., Hazelton, J.M., Williams, L.E. & Abee, C.R. (1988). *In vitro* bioactive luteinizing hormone assay shows cyclical, seasonal hormonal changes and response to luteinizing-hormone releasing hormone in the Squirrel Monkey (*Saimiri boliviensis boliviensis*). *American Journal of Primatology*, **14**, 167–75.

# 7 *Seasonality of reproductive performance in rural Gambia*

STANLEY J. ULIJASZEK

## Introduction

There are a number of ways in which the reproductive performance of women can be assessed. These include measures of fertility, pregnancy outcome (one expression of which is birthweight), lactational performance, and child mortality. Ultimately, the survival to reproductive age of as many children as possible is the most direct measure of successful reproductive performance, the other variables, including pregnancy outcome and lactational performance, being proximate factors.

Although it is acknowledged that social factors can influence fertility, these will not be addressed in this chapter. Rather, the focus is on seasonal variation in a number of factors that can cause physiological stress in women and young children. Such factors are linked to seasonality in the natural human-created agricultural environment, and for analytical purposes, can be reduced to the following: 1. food availability and intake of dietary energy; 2. energy expenditure (and also how these two factors affect energy balance); and 3. infectious disease, as it influences pregnancy outcome and child survivorship. Although it might be argued that energy balance (that is, the difference between intake and expenditure) is of prime consideration with respect to energy nutritional stress, a good case has been put forward to suggest that energy intake, expenditure (insofar as this is an expression of heavy physical activity) and balance should be considered separately in at least one aspect of reproductive performance, fertility (Rosetta, 1990). The relative importance for reproductive performance of seasonal variation in these three factors will be considered for one rural agricultural system in the Gambia.

## Types of seasonality

In populations living in tropical regions, infectious disease, and energy intake, expenditure and balance may vary seasonally, but may not all vary in the same way, or at the same time. There are societies in which individual

76

seasonal stresses operate, for example in urban Mexico, where the major seasonal stress is infection (Sepulveda *et al.*, 1988), and there are societies in which multiple stresses operate. Urban Third-World communities are more likely to fall into the first category, while rural communities are more likely to fall into the second. Of societies where multiple stresses have been shown to operate, these may or may not be coincidental. In general, agricultural communities are more likely to experience multiple, coincidental seasonal stress than are other types of community. The populations of rural Gambia experience this type of seasonal stress.

### The rural Gambian population

Seasonal effects on reproductive performance are considered using published data for one seasonal agricultural system in West Africa, as represented by three villages, Keneba, Kanton Kunda, and Manduar (Fig. 7.1). These villages are set in the West Kiang Administrative District, an area of savannah scrub and farm land, roughly 40 by 20 kilometres, bounded on three sides by brackish tidal rivers (Lewis, in press). The villages are linked to each other by dirt roads, and to urban centres by a laterite road.

People in these villages have been the subjects of extensive medical, demographic and nutritional research, which began with the work of Dr (later Sir) Ian McGregor in 1949, and which continues to the present day. The area and the particular villages were chosen because they were in what was then regarded as an isolated and backward part of the Gambia (Lewis, in press). The British Medical Research Council supported McGregor's work, and that of the Dunn Nutrition Unit (DNU), which in 1974 took over and developed the field research centre that McGregor had established in one of the villages, Keneba. By the mid-1980s, these villages could no longer be considered poor and backward relative to the rest of the Gambia. The availability of jobs and money from the DNU to the local communities served to reduce reliance on the subsistence economy, while a very high standard of primary health care, and numerous health and nutrition interventions provided by the DNU staff (Lewis, in press), were responsible for halving neonatal mortality rates, and reducing infant mortality rates between 1974 and 1982–3 from 149 to 25 per 1000 live births (Lamb *et al.*, 1984). These and other factors, such as seasonal migration for paid labour and dietary supplementation of pregnant women in the third trimester of pregnancy, may have served to reduce these communities' biological experience of seasonality. An illustration of this is given in Fig. 7.2, which shows a decline in overall seasonal weight change in women between 1978 and 1985.

The data used in this article come from a variety of published sources,

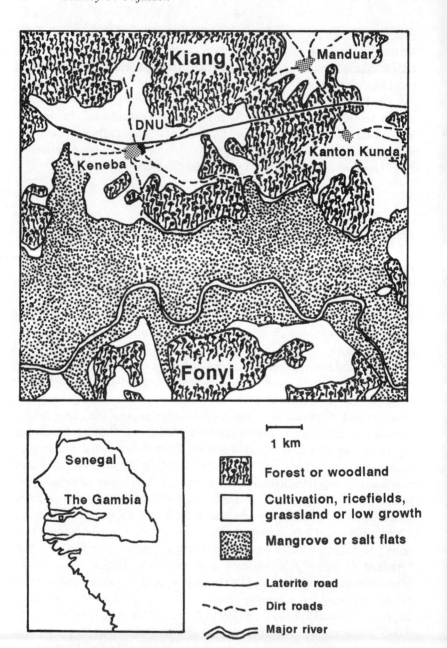

Fig. 7.1. Map of the Gambia, and the three villages studied.

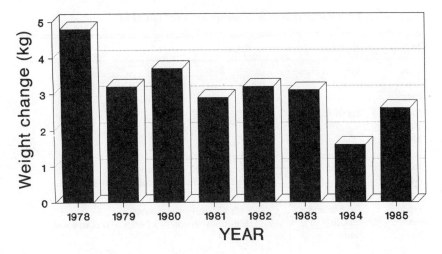

Fig. 7.2. Seasonal weight change in rural Gambian women, 1978–85. From Singh *et al.* (1989).

and were collected at different times between 1949 and 1985. Much of the published data report seasonal effects, which are statistically corrected for differences due to pregnancy and lactation (e.g. Lawrence & Whitehead, 1988; Lawrence *et al.*, 1989), and it is this type of data that is used here. Since the experience of seasonality has probably declined across time, the interpretations offered here are of necessity cautious ones.

### Climatic seasonality and seasonal stress

Table 7.1 summarises the major climatic types across the year in the Gambia, and the immunological and physiological stresses associated with them. During the period December to May of each year, there is little stress, since work output is low, food is available from the previous year's harvest, and exposure to infectious disease is low. June to August sees the start and the peak of the rainy season, a time in which food shortages are experienced, and high physical work output is required to clear and plant the fields. The transmission of malaria and diarrhoeal diseases is high at this time. During the months September to November, the rains decline then cease. Transmission of infection remains high, work output, particularly in harvesting, is high, although food becomes increasingly available with the first harvest.

Fig. 7.3 shows the proportion of time spent ill by children in Keneba village in 1974–5, illustrating the seasonality in infectious disease to be found there. Diarrhoea, lower respiratory tract infection and malaria all

Table 7.1. *Climate and seasonal stress in rural Gambia*

| Month | Weather | Stresses |
|---|---|---|
| Dec.<br>Jan.<br>Feb. | Dry and cool | |
| Mar.<br>Apr.<br>May | Increasingly hot | |
| Jun.<br>Jul.<br>Aug. | Start and<br>peak of<br>rains | Shortage of food<br>High physical work output<br>High transmission of malaria and diarrhoeal<br>diseases |
| Sep.<br>Oct.<br>Nov. | Decline and<br>cessation of<br>rains | High physical work output<br>High transmission of malaria and diarrhoeal<br>diseases |

After Rowland *et al.* (1981).

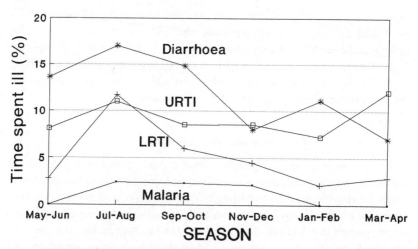

Fig. 7.3. Proportion of time spent ill by children in Keneba village, 1974–5. URTI = upper respiratory tract infections; LRTI = lower respiratory tract infections. From Rowland *et al.* (1981).

show marked seasonal prevalences, although children spend 13 times as long being ill with diarrhoea than with malaria (Rowland *et al.*, 1981).

A comparison of seasonality in energy expenditure in physical activity of women in 1949–50, and 1982–5 is given in Fig. 7.4. The data for 1949–50 come from a region of the Gambia different from that of 1982–5, so the comparison can only be suggestive. It would appear, however, that there has been a decline in seasonal variation in activity levels, mainly due to

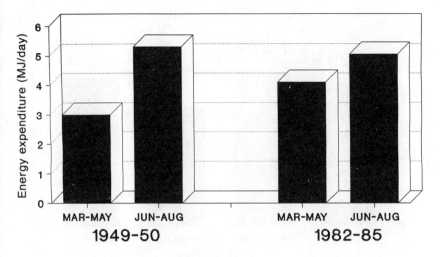

Fig. 7.4. Estimated energy expended in physical activity by rural Gambian women, 1949–50 and 1982–5. From Haswell (1953) and Lawrence & Whitehead (1988).

greater dry season energy expenditure in 1982–5 compared with 1949–50. There is little difference in wet season activity level between these two times of measurement. The difference in dry season energy expenditure at activity is in the order of 1 MJ per day, and can be attributed largely to increases in the time spent performing household maintenance activities, including food preparation, drawing and carrying water, washing clothes and dishes, and so on. In 1949–50, women would spend an average of 3.5 hours per day in such tasks (Haswell, 1953), while in 1982–5, the value reported was 4.8 hours per day (Lawrence & Whitehead, 1988). Time spent in agricultural activities showed little difference across time, for either time of year.

Other seasonal stresses coinciding in the wet season in the Gambia have been reported to be low food intake (Prentice *et al.*, 1981) and low levels of breastmilk production (Prentice *et al.*, 1983). The relative contribution of these factors to reproductive performance will be considered in the next section.

### Reproductive performance

Billewicz & McGregor (1981) demonstrated clear seasonality of birthrates, the lowest number of births taking place at the beginning of the wet season (Fig. 7.5). Although lower frequency of coitus in the second half of the wet season is possible (although not reported) it cannot explain this pattern of birth rates. If coital frequency is reduced because of tiredness due to heavy physical work, and birth rate is a function of coital frequency, then lower

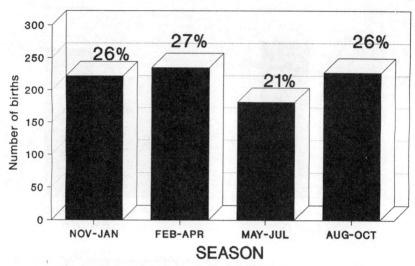

Fig. 7.5. Seasonality of births in Keneba. From Billewicz & McGregor (1981).

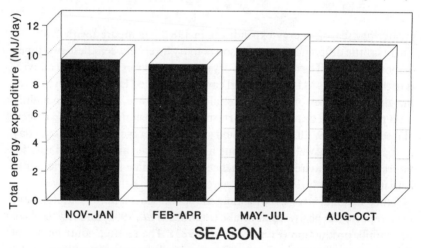

Fig. 7.6. Total daily energy expenditure of rural Gambian women by season, 1982–5. From Lawrence & Whitehead (1988).

birth rates would be expected between February and April as well as May to July, since the heaviest agricultural period comes in the early wet season, and not in the late wet season. It is more likely that the effect shown in Fig. 7.5 is due to the cumulative effects of heavy work load and negative energy balance across the wet season.

Total daily energy expenditure for pregnant and lactating women in the early wet season (May to July) is higher than at any other time of year, including the late wet season, August to October (Fig. 7.6). Energy balance

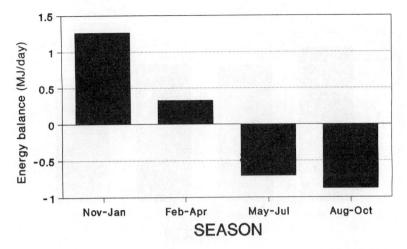

Fig. 7.7. Energy balance (estimated from body weight and skinfold thickness measures) of rural Gambian women by season. From Lawrence *et al.* (1989).

(estimated from changes in body weight and skinfold thicknesses) is positive in the dry season, negative in the wet (Fig. 7.7). The greatest positive energy balance comes in November to January (after the harvest) and the greatest negative energy balance comes between August and October, before the harvest, but not at the period of greatest energy expenditure. In this analysis, energy intake is estimated from energy balance and expenditure, and not from direct measurement of intake, since Singh *et al.* (1989) have shown that for Gambian women, accurate measures of average total energy expenditure exceed estimates of energy intake by over 50%. Using this estimate, energy intake is greatest between November and January, and lowest between August and October (Fig. 7.8). It would appear that the period of greatest negative energy balance is more closely related to low food intake than to seasonally higher energy expenditure.

When the output of breastmilk is examined, there is a steady decline with the duration of lactation, with some seasonal variation, 12-hour breastmilk output being lowest in the August to October period (Fig. 7.9). With respect to protection from ovulation by mechanisms associated with lactation, it is unlikely that these are particularly important during the wet season, if Prentice *et al.* (1986) are to be believed. They suggest that there is a strong drive towards milk synthesis in Gambian women, milk output being not limited by food intake, but controlled by the characteristics of the mother–infant pair. This view contradicts evidence from Bangladesh, where the amount of time spent suckling during the times of high work output and throughout the wet season is higher than during the dry season

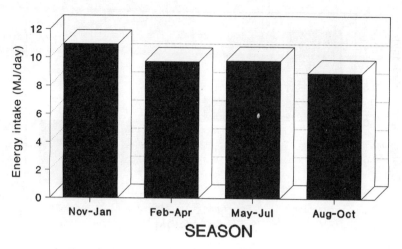

Fig. 7.8. Energy intake (estimated from energy expenditure and balance measures) of rural Gambian women by season.

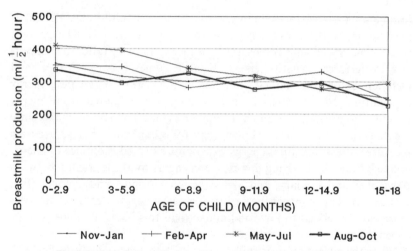

Fig. 7.9. Seasonality of breastmilk production in rural Gambian women. From Prentice *et al.* (1983).

when work output is low (Chowdhury *et al.*, 1981). Wet season–dry season differences in suckling behaviour need to be clarified before the influence of breastfeeding on wet season fertility can be determined with any confidence in the Gambia.

Women in Keneba show differences in energy balance, as estimated from measures of body composition, at all stages of pregnancy, between wet and dry seasons (Table 7.2). One result of this difference in energy balance is that birthweights are about 200–300 g lower in the wet season than in the dry (Prentice *et al.*, 1981).

Table 7.2. *Estimates of energy balance, from changes in body composition*

| | Energy balance (MJ/day) | |
|---|---|---|
| Months of pregnancy | Dry season | Wet season |
| 0–2.9 | +1.16 | −0.31 |
| 3–5.9 | +0.61 | −0.22 |
| 6–9 | +0.84 | +0.10 |

From Prentice *et al.* (1981).

One seasonal difference is that of higher mortality in early infancy, which may be related to low birthweight (less than 2.5 kg). Although the majority of low birthweight births may be attributable to the lower energy intakes experienced by women passing their third trimester of pregnancy during the wet season, infection by malaria is also important, particularly among primiparous women (Watkinson & Rushton, 1983). In Keneba, primips represent 18% of all births, but about 30% of them are affected by malarial infection *in utero*, as assessed by placental pigmentation. First-born infants with pigmented placentas due to malarial infection have mean birthweights of 2.58 kg, compared to 3.15 kg for unaffected babies. This must contribute significantly to neonatal mortality, although the extent of this has not been investigated.

### Seasonality of child survivorship

Table 7.3 gives mortality rates by season and age for rural Gambian children. Those children born between August and October, and November and January have the highest mortality rates in the first three months of life. Although three-monthly mortality rates are highest for all birth cohorts when they are in the wet season (Billewicz & McGregor, 1981), the November to January birth cohort has the highest overall mortality rate in the first year of life, while the May to July birth cohort has the lowest mortality rate for the same period. The latter cohort also has the greatest proportion of children surviving to the age of five years. Children born in May to July enter the wet season with higher average birthweights than children born later on in the wet season. Breastfeeding is likely to shield them from excessive exposure to diarrhoeal diseases for most of the wet season, the introduction of weaning foods taking place when the worst of the wet season is over. Most of the infant mortality in rural Gambia is due to the interactive effects of undernutrition and infection; this synergism is greatest in the wet season, but still operates at a lower level in the dry season when the incidences of respiratory tract infection and diarrhoea are lower if not completely absent (Rowland *et al.*, 1981).

Table 7.3. *Young-child mortality in Keneba*

| Season of birth | 0–3 month mortality rate (per 1000 live births) | Chances of dying (%) | |
|---|---|---|---|
| | | By 1 year | By 5 years |
| Nov.–Jan. | 135 | 32 | 54 |
| Feb.–Apr. | 85 | 21 | 49 |
| May–Jul. | 88 | 15 | 40 |
| Aug.–Oct. | 123 | 25 | 55 |

From Billewicz & McGregor (1981).

Comparing the mortality of children by season of birth (Table 7.3) with seasonality of fertility (Fig. 7.5), it is apparent that the birth cohort with greatest young child survivorship, May to July, is the one with the lowest proportion of births, compared with other times of the year. It would appear, therefore, that women who conceive late in the wet season have an advantage in reproductive performance compared with women who conceive at other times of the year. There must be constraints on conception at this time, since it seems logical that all women would seek to maximise their reproductive performance in this way, given the opportunity, giving rise to higher birth rates than average at the beginning of the wet season, rather than the lower rates observed.

An obvious constraint is an energetic one, of low energy intakes and negative energy balance, possibly interacting with lactation; this could lead to an impairment in ovulation by the mechanism postulated by Rosetta (1990), through blockage of the hypothalamic pulsatile generator of gonadotrophin release hormone, thereby reducing conception rates. Another constraint might be reduced coital frequency during the wet season because of physical exhaustion. Yet another could be the effect that high birth rates at the beginning of the wet season might have on work schedules.

It has been shown that in the wet season, late pregnant and early lactating women go to the fields to work for fewer days than do women who are in early pregnancy or late lactation (Lawrence & Whitehead, 1988) (Table 7.4). When the percentage of all days spent in the fields is crudely translated into days per seven-day week, women in late pregnancy or early lactation spend 4.7 and 3.5 days in the fields during the early wet and late wet seasons respectively, compared with 5.7 and 5.4 days for women in early pregnancy or late lactation. In addition, on the days when women in late pregnancy or early lactation go to the fields, they expend less energy at work than their counterparts who are not at either stage of the reproductive

Table 7.4. *Proportion of all days spent in the fields by rural Gambian women, by season*

| Season | Days spent in fields (%) | |
|---|---|---|
| | 28 weeks pregnant – 4 weeks post-partum | Early pregnancy and later lactation |
| Dry | 30 | 26 |
| Early wet | 67 | 81 |
| Late wet | 50 | 77 |
| Harvest | 31 | 42 |

From Lawrence & Whitehead (1988).

cycle. If the energy costs of pregnancy are in large part borne by this type of reduction in physical activity, there may be limits to the amount of accommodation possible. In the early wet season, late pregnant or early lactating women form about 20% of the total work force. Cooperative organisation of work may allow these women, one fifth of the female work force, to spend one day per week less in the fields in the early wet season, if the other 80% of women work one day more per month. If a greater proportion of women were to give birth at the beginning of the wet season, it might not be possible to buffer them and their babies against energetic stress without a reduction in overall agricultural work output. It can be argued that a reduction in work output could lead to a decline in food production, which could affect the nutritional well-being of the entire community. Thus the reproductive advantage of the few would be bought at the cost of nutritional disadvantage to the entire community.

### Conclusions

In this chapter the relationships that exist between birthrate, pregnancy outcome, lactational performance and child mortality are examined in a rural community in the Gambia, where seasonality of food intake, energy expenditure and infectious disease are experienced. Seasonal troughs in birthrate are more closely related to seasonally low intakes of dietary energy, and childhood mortality is lowest in children conceived at the time of year when fertility is lowest.

### Acknowledgements

I thank Dr S. Poppitt, Professor P. Shetty and Dr S. Strickland for helpful comments on earlier drafts of this chapter.

88    *Stanley J. Ulijaszek*

## References

Billewicz, W.Z. & McGregor, I.A. (1981). The demography of two West African (Gambian) villages, 1951–75. *Journal of Biosocial Science*, **13**, 219–40.

Chowdhury, A.K.M.A., Huffman, S.L. & Chen, L.C. (1981). Agriculture and nutrition in Matlab Thana, Bangladesh. In *Seasonal Dimensions to Rural Poverty*, ed. R. Chambers, R. Longhurst & A. Pacey, pp. 52–61. London: Frances Pinter.

Haswell, M.R. (1953). Economics of agriculture in a savannah village. *Colonial Research Study No. 8*. London: Her Majesty's Stationery Office.

Lamb, W.H., Foord, F., Lamb, C. & Whitehead, R.G. (1984). Changes in maternal and child mortality rates in three isolated Gambian villages over ten years. *Lancet*, **ii**, 912–14.

Lawrence, M. & Whitehead, R.G. (1988). Physical activity and total energy expenditure of child-bearing Gambian village women. *European Journal of Clinical Nutrition*, **42**, 145–60.

Lawrence, M., Lawrence, F., Cole, T.J., Coward, W.A., Singh, J. & Whitehead, R.G. (1989). Seasonal pattern of activity and its nutritional consequence in Gambia. In *Seasonal Variability in Third World Agriculture*, ed. D.E. Sahn, pp. 47–56. Baltimore: Johns Hopkins University Press.

Lewis, G.A. (in press). Some studies of social causes and cultural responses to disease. In *The Anthropology of Disease*, ed. C.G.N. Mascie-Taylor. Oxford: Oxford University Press.

Prentice, A.M., Whitehead, R.G., Roberts, S.B. & Paul, A.A. (1981). Long-term energy balance in child-bearing Gambian women. *American Journal of Clinical Nutrition*, **34**, 2790–9.

Prentice, A.M., Roberts, S.B., Prentice, A., Paul, A.A., Watkinson, M., Watkinson, A.A. & Whitehead, R.G. (1983). Dietary supplementation of lactating Gambian women. I. Effect on breast-milk volume and quality. *Human Nutrition: Clinical Nutrition*, **37C**, 53–64.

Prentice, A.M., Paul, A.A., Prentice, A., Black, A., Cole, T.T. & Whitehead, R.G. (1986). Cross-cultural differences in lactational performance. In *Human Lactation. 2. Methods and Environmental Factors*, ed. M. Hamosh & A.S. Goldman, pp. 13–44. New York: Plenum.

Rosetta, L. (1990). Biological aspects of fertility among Third World populations. In *Fertility and Resources*, ed. J. Landers & V. Reynolds, pp. 18–34. Cambridge: Cambridge University Press.

Rowland, M.G.M., Paul, A., Prentice, A.M., Muller, E., Hutton, M., Barrell, R.A.E. & Whitehead, R.G. (1981). Seasonality and the growth of infants in a Gambian village. In *Seasonal Dimensions to Rural Poverty*, ed. R. Chambers, R. Longhurst & A. Pacey, pp. 164–75. London: Frances Pinter.

Sepulveda, J., Willett, W. & Munoz, A. (1988). Malnutrition and diarrhea. A longitudinal study among urban Mexican children. *American Journal of Clinical Nutrition*, **127**, 365–76.

Singh, J., Prentice, A.M., Diaz, E., Coward, A.W., Ashford, J., Sawyer, M. and Whitehead, R.G. (1989). Energy expenditure of Gambian women during peak agricultural activity measured by the doubly-labelled water method. *British Journal of Nutrition*, **62**, 315–29.

Watkinson, M. & Rushton, D.I. (1983). Plasmodial pigmentation of placenta and outcome of pregnancy in West African mothers. *British Medical Journal*, **287**, 251–4.

# 8    Seasonal effects on physical growth and development

T.J. COLE

As auxologists know, le Comte de Montbeillard measured his son's height every six months between 1759 and 1777. Buffon, who reported the findings (Tanner, 1981), noticed that the son grew twice as much in the summer as in the winter; this was probably the first description of a seasonal component to growth. The subject has remained a firm favourite among auxologists ever since, with a substantial literature dating back more than a century (e.g. Malling-Hansen, 1886; Orr & Clark, 1930). The early studies were European, but North Americans made an increasing impact from the beginning of this century (E.L. Marshall, 1937), and seasonality of growth in the developing world has been documented since the middle of this century (see Valverde *et al.*, 1972, for a comprehensive review).

Seasonal trends in growth are very different in the developed world and the developing world, so it is important to distinguish between them. In the developing world growth rate is very clearly related to climatic factors associated with the timing of the rainy season or seasons, through their influence on food availability, parasite load (e.g. malaria) and infection. The absence of a rainy season in most parts of the developed world means that the most obvious seasonal influences do not apply, so that seasonality of growth is much more subtle. However, this is not to say that factors such as infection have no effect on growth in the developed world – they are just harder to identify.

In the developed world, the broad consensus is that (a) height velocity is greatest in the spring (W.A. Marshall, 1971), (b) weight velocity is greatest in the autumn (e.g. E.L. Marshall, 1937), (c) these effects are less obvious in infancy than in childhood, and (d) the effects are very variable, with only a minority of individuals following the mean trend. However, many of the relevant studies have had methodological problems, in that the measurements are widely spaced across the year, or else the analysis is inadequate, with insufficient control for confounding factors such as age. Berkson (1930) went to the opposite extreme, using a sophisticated analysis to fit quadratic curves to the mean weight and length curves of 15 000 children, and he found a seasonal pattern in the residuals. Unfortunately, he omitted to record the season when this pattern occurred!

89

The aims of this chapter are to examine in detail the seasonal nature of infant growth in two contrasting areas of the world, the Gambia and England; to see how well it matches the pattern described in the literature, and to judge its impact on the routine assessment of growth. The issue of seasonality in birthweight and pregnancy weight gain is not covered here, as it has already been discussed extensively, in the context of the Gambia, by Prentice *et al.* (1987). However, the large database of maternal weights available in the Gambia is used to provide a perspective on the nature of seasonal weight change as it affects the whole community.

## Subjects

The Gambian data are of mothers and children from three rural villages (Keneba, Kanton Kunda and Manduar) in West Kiang (see Fig. 7.1). The villages have been studied continuously by the Dunn Nutrition Unit since 1974 (Rowland *et al.*, 1981; Prentice *et al.*, 1987). The seasonal pattern in the Gambia revolves around a single rainy season starting at the end of May. This is the time when food stocks from the previous year are running out, and the adults prepare the land for sowing. As soon as the rains come the crops are planted. In a good year the rains last for up to two months, but in a bad year they may be virtually absent. The first crops are harvested during September, and other crops continue to appear until November. Thus from June to September adults are required to work hard in the fields while their food stocks are dwindling. This period is known as the 'hungry season'. From September to December the humidity falls, and then from December to May the following year the temperature continues to rise.

Data collection in the Gambia has occurred at regular morning clinics to which the subjects are invited, initially four-weekly and later six-weekly. Weight is recorded for women, and both weight and length are recorded for children up to two years of age. Women usually attend the clinic throughout pregnancy and lactation. The child data used in the present analysis were collected between May 1974 and April 1980, while the maternal data extended from April 1978 to the end of 1988. During this period women had repeated pregnancies, up to a maximum of five. Lactating women in Keneba received a dietary supplement from May 1979 (Prentice *et al.*, 1983a), while a maternal pregnancy supplement was started in May 1980 (Prentice *et al.*, 1987). The stage of pregnancy or lactation, used for adjustment in the regression analyses, is defined relative to the date of delivery. Exact dates of birth are available for all the children and many of the mothers.

The Cambridge data for analysis consist of infants recruited into the Cambridge Infant Growth Study between 1984 and 1988 (Whitehead *et al.*,

1989). Infants were measured in their own homes every four weeks from four to 52 weeks of age, plus or minus two days. Measurements included weight, length, arm circumference, head circumference, triceps and subscapular skinfold. Standard anthropometry methods were used in both the Gambia and Cambridge.

### Statistical analysis

The Gambian children were split into six groups according to their month of birth, for example December–January, February–March, and so on. Using linear interpolation, values of weight and length were obtained for each child at each exact calendar month of age from one to 24 months, and these were averaged by the method of Tanner (1951). This provides group mean adjusted values and increments at each month of age, making use of mean values of weight and length in previous months and the correlations between them, to improve precision. The need for each individual's measurements to be all in the same group requires the grouping to be by month of birth rather than month of measurement.

The data for many of the Gambian mothers extended over 11 years, representing several pregnancies. To exploit this structure, within-subject regression analysis was used. This is ordinary least-squares regression, but an individual constant term is fitted for each mother in the model, and so each mother acts as her own baseline. However, the constants are not fitted as regression coefficients; instead the means of the data for each mother are subtracted from her data at the start of the analysis. This reduces the complexity of the regression analysis materially, as otherwise the constant terms would need to be estimated for each of the several hundred subjects. One side-effect of the analysis is that variables that are constant for each mother, such as height, cannot be included in the model.

The data for the Cambridge infants were also analysed using a form of within-subject regression. Their six anthropometry measures were converted to standard deviation scores ($Z$ scores) using the LMS method (Cole, 1990), and quadratics in log(weeks) were fitted to each child's six sets of $Z$ scores. This was done to remove the effects of catch-up or catch-down on $Z$ score, using a log time scale because such trends tend to occur early in life. The residuals from the fits were then analysed by within-subject regression. As a subsidiary analysis, the change in $Z$ score from one occasion to the next for each child was modelled.

To look for seasonal effects in Cambridge, and taking note of the measurements being made every four weeks, the infants were divided up into 13 groups according to their lunar month of birth (where lunar month 1 is 1st–28th January, and so on). This ensures that for example infants

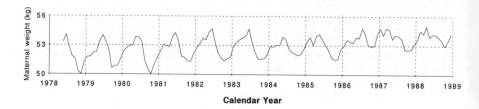

Fig. 8.1. Seasonal weight change in 529 Gambian women between 1978 and 1988. The data are adjusted for stage of pregnancy/lactation using within-subject regression.

aged four weeks (or one lunar month) and born in lunar month 6 (21st May to 17th June) are measured in lunar month 7. The same holds for children 6 lunar months old measured in lunar month 1, and equally for other children whose sum of age and month of birth is equal to seven. In this way age and month of birth exactly specify the month of measurement. Each of these three factors is adjusted for the other two in the regression analysis.

### Results

#### Gambian mothers

Over the 11-year period 1978–88, 13 833 maternal weights were collected (mean 52.8 kg, SD 7.6) on 529 women, a mean of 26 (median 17) weights per woman (range one to 94). For 501 weights, the mother's stage of pregnancy and/or lactation was not known. Within-subject regression was used to adjust for stage of pregnancy/lactation, year of measurement and calendar month of measurement, assuming an additive model. The residual weight SD was reduced to 2.1 kg after fitting the model.

Fig. 8.1 shows weight adjusted for stage of pregnancy/lactation and plotted against calendar month and year for the period 1978–88. There is a steep fall in weight over the hungry season every year, and also a consistent dip in April, which has not so far been explained. The most obvious change over the period, apart from the upward trend, is the reduction in hungry season weight loss from 4 kg in 1978 to 2 kg in 1988.

Figs. 8.2 to 8.4 show the mean effects of stage of pregnancy, year and month for the model, where the standard errors for differences between points in each figure are 0.15, 0.1 and 0.1 kg respectively. In Fig. 8.2 the mean weight gain throughout pregnancy is nearly 8 kg, well below the recommended value of 12.5 kg. This is largely due to the seasonal fall in weight during the hungry season, which affects most pregnancies (Prentice *et al.*, 1987). Fig. 8.2 also shows a slight rise in weight over the early months of lactation.

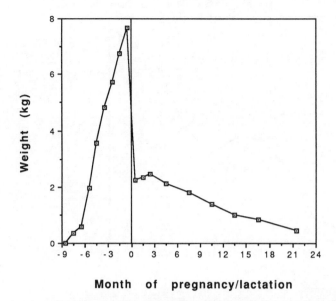

**Month of pregnancy/lactation**

Fig. 8.2. Weight gain and weight loss during pregnancy and lactation in Gambian women, expressed relative to the pre-pregnant state. Each point has a standard error of about 0.15 kg. The data are adjusted for calendar month and year of measurement using within-subject regression.

The year effect on weight, shown in Fig. 8.3, is a fairly linear increase over the 11 years, so that on average women are 2 kg heavier in 1988 than in 1978. A separate analysis to test for trend in weight with maternal age between women failed to show any significant age trend. Thus the year effect is a secular trend that has affected all women over the period of study, rather than simply a cross-sectional difference between older and younger women.

Fig. 8.4 shows the average seasonal change in weight by calendar month. The average Gambian woman loses 2.6 kg during the hungry season each year, representing 5% of her body weight. The inconsistency in weight between April and May is also obvious, which could be interpreted either as faltering during March and April, or else as a sudden increase in May and June. No adequate explanation for this feature has yet been proposed.

Taken together, the reducing hungry season loss and increasing trend in weight (Fig. 8.3) indicate significant improvements in nutritional status. Possible reasons for this are many, including better health, increased wealth, and improved sanitation. The Dunn Nutrition Unit has provided a medical service to the villages from 1974 to date, and has reduced substantially the infant mortality rate (Lamb *et al.*, 1984). Also wealth has increased throughout the period of study, through improved local

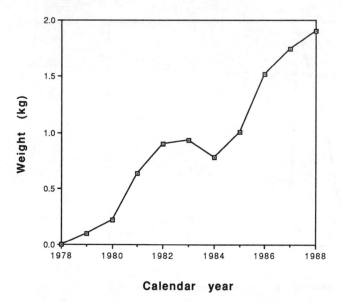

Fig. 8.3. The trend in mean weight over 11 years in Gambian women, expressed relative to 1978. Each point has a standard error of about 0.1 kg. The data are adjusted for calendar month of measurement and stage of pregnancy/lactation using within-subject regression.

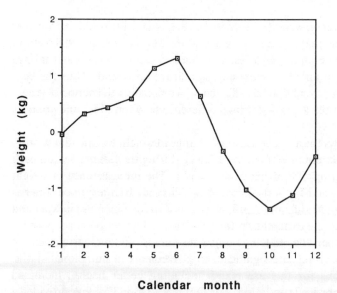

Fig. 8.4. Seasonal weight change by calendar month in Gambian women, expressed relative to the annual mean. Each point has a standard error of about 0.1 kg. The data are adjusted for year of measurement and stage of pregnancy/lactation using within-subject regression.

employment opportunities and greater contact with the outside world. This contact was substantially increased at the end of 1981, when a new road was built connecting Keneba to the Gambian Highway.

Improved sanitation is another possible explanation for the increase in weight. During 1985 standpipes were provided in Keneba, supplying good quality water from bore holes. They replaced the village well-water of often dubious quality that women had had to use before. This change in water supply had two main effects: an inevitable reduction in the energy expended by women getting water (turning a tap is less effort than lifting a bucket 30 m), and the *possibility* of a reduction in water-borne disease. The former effect is much the more likely to have had an influence on weight, and the 1986 upturn in weight seen in Fig. 8.3 may well reflect this influence. However, it should be stressed that, because so many factors are involved, this can only be speculation.

An additional component to seasonal weight variation is that provided by the fast of Ramadan. Inhabitants of The Gambia are mainly Muslim, and so observe Ramadan each year. The timing of Ramadan depends on the new moon, and lasts from the appearance of one new moon to the next, a period of 28 or 29 days. During Ramadan, Muslims may not eat or drink during the hours of daylight, but they can make up for it before dawn and after dusk. Pregnant and lactating women can in principle postpone their fast until a more convenient time, but in practice they find it easier to fast at the same time as everyone else. In temperate climates the effect of Ramadan on nutritional status in healthy subjects is minimal (Rashed, 1992), but in The Gambia during the rainy season, Ramadan has a substantial effect on the health of pregnant women (Prentice *et al.*, 1983b).

Each year Ramadan starts about 11 days earlier than the year before, so that between 1978 and 1988 it moved back from 5th August to 17th April. Because of this, it exerts a seasonal influence on weight over and above that of the calendar month, and the two effects can be disentangled using within-subject regression. This is only possible for studies lasting several years. In order to establish the weight changes occurring around Ramadan, weight was adjusted for stage of pregnancy, year and month as described above, but in addition separate weight effects were obtained for each of the 12 weeks bracketing Ramadan, that is, four weeks before, four weeks during and four weeks after. Each effect was expressed relative to weight at other times of the year, that is, between successive Ramadans.

Fig. 8.5 shows how weight changes in the 12 weeks around Ramadan, adjusted for other factors. The numbers of weights in each week range between 150 and 350, providing standard errors for each week of about 0.15 kg. The figures shows that before Ramadan, weight is consistently about 0.3 kg above the baseline, while afterwards it is 0.6 kg below. At the

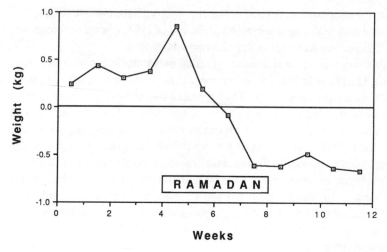

Fig. 8.5. Weight change in Gambian women during 3 months around the fast of Ramadan, expressed relative to mean weight for the other 9 months of the year. Each point has a standard error of about 0.15 kg. The data are adjusted for calendar month and year of measurement, and stage of pregnancy/lactation, using within-subject regression.

start of Ramadan weight increases appreciably, probably because the pre-dawn breakfast is larger than before. Weight then falls, approximately linearly, over the 4 weeks of Ramadan, by a total of 1.5 kg. The end of Ramadan is usually celebrated with a feast, and this probably cancels out the effect of breakfast reverting to its pre-Ramadan size.

The acute effect of Ramadan on the weight of these women is thus 1.5 kg, while the chronic drop is 0.9 kg, from $+0.3$ to $-0.6$ kg. The deficit is made up during the rest of the year, so that by the time Ramadan comes round again weight is back to where it was before. However this may be an incomplete view, as the 11 years of the study covered Ramadan only during the period April to August, a relatively strenuous time of year. To get a more balanced picture, including Ramadan over the whole year, would require 33 years of data. As it is, the observed effect is applicable to Ramadan when it occurs around the time of the rainy season.

### Gambian children

A total of 686 Gambian children, 351 female, contributed 5814 measurements to the weight and length analyses, providing between 26 and 119 measurements at each month of age in each birth date group. The results for mean weight and length are shown plotted in Figs. 8.6 and 8.7 respectively, with the growth curves for each of the six birth groups shown

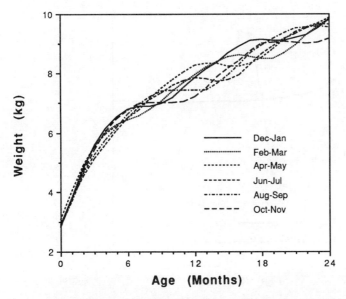

Fig. 8.6. Weight charts for Gambian children from birth to 24 months of age, grouped by month of birth. The data are adjusted by the method of Tanner (1951).

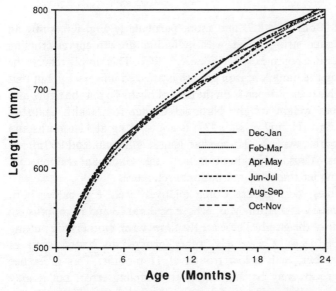

Fig. 8.7. Length charts for Gambian children from 1 to 24 months of age, grouped by month of birth. The data are adjusted by the method of Tanner (1951).

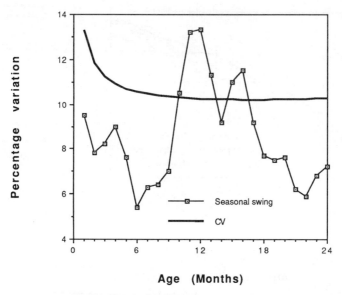

Fig. 8.8. The percentage variation in Gambian mean weight attributable to season (lower line) compared with the population coefficient of variation (CV) of weight for Cambridge infants, by month of age.

superimposed. Both Fig. 8.7, and more particularly Fig. 8.6, show an interesting 'plaited string' effect, with individual growth curves crossing over each other in a complex way (Billewicz, 1967). This simply reflects the fact that during the hungry season growth is affected universally, but that the age when this occurs depends on the date of birth. To put the figure into context, median weight on the National Center for Health Statistics (NCHS) standard (Hamill *et al.*, 1977) is about 10 kg at 12 months and 12 kg at 24 months, while NCHS median length is 750 mm and 860 mm at the same ages. Thus, throughout infancy the Gambian children are appreciably smaller than the international reference.

Fig. 8.6 shows some groups doing relatively well, e.g. babies born December–January and April–May, whose peaks at 13 and 17 months are on the high side of the graph. These are the times when mothers are putting on weight (Fig. 8.4). Conversely, other groups do badly, such as October–November, with a low trough at 11 months. This difference between the groups may be due simply to sampling error, but it may alternatively be an indication that certain months of the year are better birth months than others.

The magnitude of the seasonal swing at each age can be summarised by expressing the observed range as a percentage of the mean value. For weight (Fig. 8.8) the seasonal swing is 9% at one month, down to 6% at

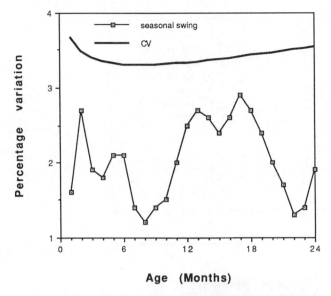

Fig. 8.9. The percentage variation in Gambian mean length attributable to season (lower line) compared to the population coefficient of variation (CV) of length for Cambridge infants, by month of age.

seven months of age, rising steeply to 13% at 11 months, and gradually falling back to 7% by 21 months. For comparison Fig. 8.8 shows the coefficient of variation (CV) of weight by age in Cambridge infants (Cole, 1988). The CV is about 10% of the mean at 12 months, so that the seasonal swing represents more than a unit change in $Z$ score.

For length a similar pattern is seen, except that the swing is much smaller (Fig. 8.9). It ranges between 1.3%, occurring at eight and 22 months, and 2.6%, between 12 and 18 months. The CV of length in Cambridge infants by age is also shown in Fig. 8.9. At 12 months it is about 3.5%, so that the seasonal swing is about 0.75 $Z$ score units, rather less than for weight.

Figs. 8.10 and 8.11 show weight velocity and length velocity for the same children grouped by birth date, again using the efficient estimates provided by the method of Tanner (1951), which adjusts the group mean velocities for mean weight and length in previous months. The figures confirm that seasonality of growth, even in this harsh environment, is hardly evident before three months of age; over this period velocity falls steeply in all the groups. After nine months there is only a weak trend in velocity with age, but strikingly in every group mean weight velocity is negative at its lowest point.

The peaks and troughs of velocity in Figs. 8.10 and 8.11 are much more consistent from group to group than the corresponding peaks and troughs

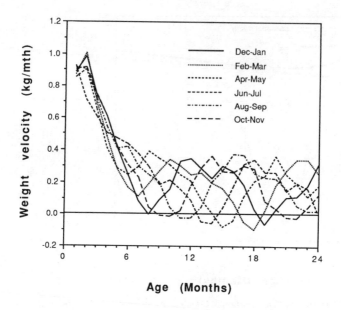

Fig. 8.10. Weight velocity charts for Gambian children from 1 to 24 months of age, grouped by month of birth. The data are adjusted by the method of Tanner (1951).

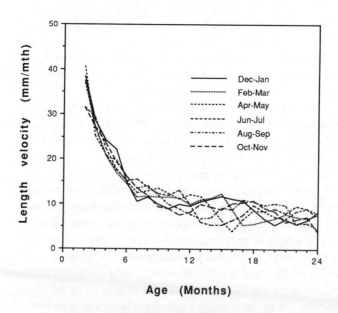

Fig. 8.11. Length velocity charts for Gambian children from 1 to 24 months of age, grouped by month of birth. The data are adjusted by the method of Tanner (1951).

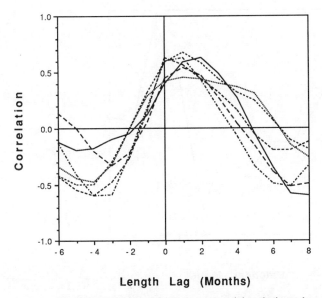

Fig. 8.12. Correlations between mean weight velocity and mean length velocity, in each of six birth-month groups, calculated over the age range 7–24 months. The correlations are computed for a range of time intervals (lags) between the timing of the weight velocity and the corresponding length velocity. See text for details.

for distance attained (using Tanner's nomenclature) in Figs. 8.6 and 8.7, although the troughs are rather lower at 12 months than at other ages. The group differences in distance are likely to be due to influences operating very early in life, as summarised by birthweight and growth in the first three months.

Examination of Figs. 8.10 and 8.11 suggests that the peaks and troughs in weight velocity occur earlier than the corresponding extremes for length velocity. This is confirmed by calculating lagged correlations between weight velocity and length velocity for each group, excluding the first six months of age. The results in Fig. 8.12 show that the strongest correlations are obtained for a lag of one to two months. Thus weight leads length by about six weeks in its seasonal fluctuation. This is not too surprising, since weight is a much more labile measurement than length, and unlike length its velocity can become negative.

*Gambian mothers between-subject analysis*

The seasonal variation in maternal weight gain during pregnancy has important implications for birthweight (Prentice *et al.*, 1987). Infant growth is also strongly seasonal, as shown here, and this may correspond to a critical period of growth and may retain its influence into adulthood

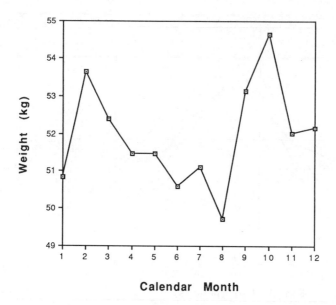

Fig. 8.13. Mean weights of 248 Gambian women by calendar month of birth. Each point has a standard error of about 2 kg. The data are adjusted for calendar month and year of measurement, and stage of pregnancy/lactation.

(Barker *et al.*, 1989). Thus it is reasonable to ask whether there is an optimal time of year to be born.

The within-mother analysis was used to calculate a mean weight for each mother, adjusted for stage of pregnancy/lactation, year and month of measurement, and Ramadan. Each mother was also given a weighting corresponding to the number of weights that had been averaged, ranging from one to 94. The exact dates of birth were known for 248 mothers, and their weights were compared according to their month and year of birth. There was no significant trend in weight with year of birth, so it was omitted from further analysis. Fig. 8.13 shows the mean weights of women by month of birth, where each point has a standard error of about 2 kg. The months April to August are consistently lower than other months, and the mean weight for these months is significantly less than for the rest of the year (difference 1.9, SE 0.8 kg, $t = 2.3$, $P < 0.05$). Repeating the analysis unweighted on the smaller number of women with known height ($n = 217$) showed those born between April and August to be shorter than the others (deficit 1.9, SE 0.8 cm, $t = 2.4$, $P < 0.05$). However, there was no evidence of a difference in body mass index (weight/height$^2$) between the two groups.

Weight and height measurements were also available for a small group of young men measured in February 1987 ($n = 62$, age range 15 to 38 years).

However, repeating this analysis including an age adjustment failed to show any significant differences attributable to their month of birth.

Overall, there is some evidence to support the hypothesis that birth during certain times of the year, specifically September to March, is better than at other times for subsequent growth and development. However, against this, the significance level of 5% is weak, the men fail to show the effect, and the hypothesis being tested, i.e. the choice of months, has been determined from the data rather than *a priori*. For all these reasons the evidence can be viewed only as suggestive.

### Cambridge infants

Turning to the developed world, data were available for 203 Cambridge infants at four weeks of age, reducing to 191 by 52 weeks, a total of 2518 measurement occasions. The subjects were recruited in three cohorts, so that their dates of birth were not uniformly distributed throughout the year. In particular only seven infants were born in lunar months 7 to 10 (July to September).

After converting all measurements to $Z$ scores, and fitting quadratic curves in age to each subject's sets of data, the residuals were analysed by regression. The aim was to see if measurements taken at particular times of the year tended to lie consistently above (or below) the subject's fitted curve. The output from the regression, for each measurement, was a set of regression coefficients consisting of a $Z$ score increment for each lunar month of measurement, representing the mean residual for that month. The significance of each coefficient could be expressed as a $t$ value, by dividing it by the standard error. Large values of the coefficient and its $t$ value would be suggestive of a seasonal trend in growth.

The residual variation in the six measurements, after adjusting for individual trends, ranged between 0.19 $Z$ score units for head and 0.20 for weight, up to 0.41 and 0.55 for subscapular and triceps skinfold respectively. Thus skinfolds are much more variable from occasion to occasion than other measurements, as would be expected. The significance of the seasonal effect for each measurement is summarised by the variance ratio, on 12 and 2335 degrees of freedom. This should exceed 2.2 for significance at 5%, or 2.7 for 1% significance. The only measurements to exceed the latter figure were length and triceps skinfold (variance ratios 3.5 and 4.5 respectively), which are thus highly significant.

The individual month effects for each measurement are shown in Fig. 8.14 as $t$ values, so that effects exceeding $\pm 2$ are significant. It can be seen that only length and triceps skinfold achieve this at all convincingly, with

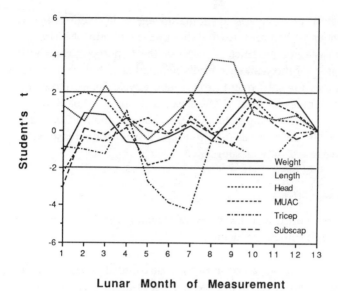

Fig. 8.14. The significance of seasonal changes in six anthropometry measures for Cambridge infants during the first year of life. The data are *t* statistics for the significance of mean residual *Z* scores by lunar month of measurement, after adjusting the *Z* scores of individuals for quadratic trends in age. MUAC = mid upper-arm circumference. See text for details.

the *t* values for length at eight and nine months reaching 3.5, and the skinfold effects at six and seven months falling to −4. The *actual* effects in the two months are +0.1 *Z* score units for length and −0.25 *Z* score units for triceps skinfold, the latter being larger due to its larger residual standard deviation. The coefficient of variation for length is about 3.5% (see above), while for triceps skinfold in these infants the CV is 21%. Thus the seasonal length spurt in lunar months 8 and 9 (August) is about 0.35% or 2 mm, averaged over the age range. The corresponding deficit in triceps skinfold for months 6 and 7 (June–July) is about 5% or 0.33 mm.

By comparison with the Gambia, these seasonal effects are tiny. Nevertheless, they are likely to be genuine, for two reasons: firstly they emerge from the regression analysis as being highly significant, and secondly the significant months are contiguous for each measurement, months 6–7 for triceps and months 8–9 for length. Fig. 8.14 shows that the two effects fit into a wider continuum, with length increasing and then decreasing between month 5 and month 13, while triceps falls then rises between months 4 and 8. This shows that the seasonal spurt in length actually occurs between months 6 and 8 (June–July), while the deficit in triceps accumulates over months 4–6 (May–June).

An alternative analysis used the change in *Z* score over each four-week period for each child, rather than the residual after fitting a quadratic trend. Unexpectedly, this gave results that were quite different from those described above. In particular, there were no significant seasonal effects at all for any of the measurements.

### Conclusions

The existence of seasonality in growth during infancy has been demonstrated in two very different parts of the world. In Cambridge, England, the effect is tiny and restricted to length and triceps skinfold. Triceps is at a minimum during May and June, while length is maximal during June and July. This agrees roughly with the pattern described in the literature, of length gain peaking in late spring and changing out of phase with weight (interpreting triceps as a proxy for weight in this instance). However, previously published results have applied more to childhood than infancy.

In the Gambia the seasonal effect is very substantial, and is synchronised with the hungry season that occurs between June and October. The percentage swing in weight-for-age from the best to the worst time of year amounts to 13% between 12 and 18 months of age. This represents a swing of well over one unit of $Z$ score, and hence is critically important in the routine assessment of nutritional status. A less extreme pattern is seen with length, with the seasonal swing amounting at its worst to 0.75 $Z$ score units. Even so, it represents a substantial component of variation over and above the population between-subject variance, and greatly reduces the value of conventional weight-for-age or length-for-age charts. There is a strong case for constructing season-specific standards for use in areas of the world where seasonal variation is of the magnitude seen in the Gambia.

Unlike Cambridge and the rest of the developed world, weight gain and length gain in the Gambia change more or less in phase with each other. There is evidence that length lags weight by some six weeks, but it is clear that both measurements respond to the same stimulus, and the only reason for the phase shift is because length, by its nature, is slower than weight to respond to the stimulus.

The large dataset of Gambian maternal weights over 11 years graphically demonstrates the reality of the annual weight loss during the hungry season. The net effect, averaged over the study, is a fall of 2.6 kg between June and October. Superimposed on this is an additional effect of 0.9 kg loss attributable to Ramadan, amounting to 3 kg loss overall when Ramadan falls in the hungry season. Over the 11 years, the seasonal weight loss has approximately halved, due probably to increasing affluence in the villages.

## References

Barker, D.J.P., Winter, P.D., Osmond, C., Margetts, B. & Simmonds, S.J. (1989). Weight in infancy and death from ischaemic heart disease. *Lancet*, **ii**, 577–80.

Berkson, J. (1930). Evidence of a seasonal cycle in human growth. *Human Biology*, **2**, 523–6.

Billewicz, W.Z. (1967). A note on body weight measurements and seasonal variation. *Human Biology*, **39**, 241–50.

Cole, T.J. (1988). Fitting smoothed centile curves to reference data. *Journal of the Royal Statistical Society, Series A*, **151**, 385–418.

Cole, T.J. (1990). The LMS method for constructing normalized growth standards. *European Journal of Clinical Nutrition*, **44**, 45–60.

Hamill, P.V.V., Drizd, T.A., Johnson, C.L., Reed, R.B. & Roche, A.F. (1977). *NCHS Growth Curves for Children Birth–18 Years*. National Center for Health Statistics, Vital and Health Statistics, Series 11: No. 165. Washington DC.

Lamb, W.H., Foord, F.A., Lamb, C.M.B. & Whitehead, R.G. (1984). Changes in maternal and child mortality rates in three isolated Gambian villages over ten years. *Lancet*, **i**, 912–14.

Malling-Hansen, P.R. (1886). Perioden im Gewicht der Kinder und in der Sonnenwärme. Kopenhagen.

Marshall, E.L. (1937). A review of American research on seasonal variation in stature and body weight. *Journal of Pediatrics*, **10**, 819–31.

Marshall, W.A. (1971). Evaluation of growth rate in height over periods of less than one year. *Archives of Disease in Childhood*, **46**, 414–20.

Orr, J.B. & Clark, M.L. (1930). Seasonal variation in the growth of school-children. *Lancet*, **ii**, 365–7.

Prentice, A.M., Roberts, S.B., Prentice, A., Paul, A.A., Watkinson, M., Watkinson, A.A. & Whitehead, R.G. (1983a). Dietary supplementation of lactating Gambian women. I. Effect on breast-milk volume and quality. *Human Nutrition Clinical Nutrition*, **37C**, 53–64.

Prentice, A.M., Prentice, A., Lamb, W.H., Lunn, P.G. & Austin, S. (1983b). Metabolic consequences of fasting during Ramadan in pregnant and lactating women. *Human Nutrition Clinical Nutrition*, **37C**, 283–94.

Prentice, A.M., Cole, T.J., Foord, F.A., Lamb, W.H. & Whitehead, R.G. (1987). Increased birthweight after prenatal dietary supplementation of rural African women. *American Journal of Clinical Nutrition*, **46**, 912–25.

Rashed, A.H. (1992). The fast of Ramadan. *British Medical Journal*, **304**, 521–2.

Rowland, M.G.M., Paul, A., Prentice, A.M., Müller, E., Hutton, M., Barrell, R.A.E. & Whitehead, R.G. (1981). Seasonality and the growth of infants in a Gambian village. In *Seasonal Dimensions to Rural Poverty*, ed. R. Chambers, R. Longhurst & A. Pacey, pp. 164–75. London: Francis Pinter.

Tanner, J.M. (1951). Some notes on the recording of growth data. *Human Biology*, **23**, 93–159.

Tanner, J.M. (1981). *A History of the Study of Human Growth*. Cambridge: Cambridge University Press.

Valverde, V., Delgado, H., Martorell, R. Belizán, J.M., Mejía-Pivaral, V. & Klein, R.E. (1972). Seasonality and nutritional status. *Archivos Latino Americanos de Nutricion*, **32**, 521–40.

Whitehead, R.G., Paul, A.A. & Cole, T.J. (1989). Diet and the growth of healthy infants. *Journal of Human Nutrition and Dietetics*, **2**, 73–84.

# 9     *Seasonal variation in the birth prevalence of polygenic multifactorial diseases*

A.H. BITTLES AND L. SANZ

## Introduction

There have been a number of reports that both single gene disorders and gross chromosomal anomalies are subject to seasonality in their birth prevalence. For example, a bimodal birth pattern was described for the autosomal recessive disorder cystic fibrosis, in England and Wales (Brackenridge, 1980a), and in Australia (Brackenridge, 1980b). Conversely, maternal age-independent chromosomal non-disjunction leading to trisomy 21 (Down syndrome) and sex chromosome aneuploidies appeared to be more epidemic in nature, with no set pattern from year to year in the months of peak prevalence (Goad *et al.*, 1976; Videbech & Nielsen, 1984). Although it is difficult to envisage specific mechanisms governing the birth frequencies of these disorders, by definition the inheritance of polygenic multifactorial disorders is subject to strong non-genetic influences (Carter, 1976) and so might be expected to exhibit seasonal variation in birth frequency. To investigate this possibility, two clinically important but very different examples were chosen for the present study. Neural tube defects (NTD), which are apparent by the end of the first month of antenatal development have been estimated to affect between 0.5 to 8.7 per 1000 livebirths worldwide (Elwood & Elwood, 1980), and schizophrenia, a disorder diagnosed in approximately 1.0% of the populations of developed countries with onset typically in adult life (McGue *et al.*, 1985).

## Seasonal variation in births

Before examining the evidence for seasonality in the births of individuals with either disorder, it is necessary to determine whether any significant variation is discernable in the overall pattern of births at different times of year. This was investigated by reference to data from England and Wales

107

for the period 1980–9, obtained from the Office of Population Censuses and Surveys (OPCS, 1990).

During the study period there were 6.562 million births in England and Wales, with a secondary sex ratio, i.e., the proportion of males to females at birth, of 105.2. As shown in Fig. 9.1, total births plotted by quarter, there was a remarkably consistent seasonality pattern across the decade, both in years when the birth rate was declining (1980–3) and increasing (1984–9). In nine of the ten years studied the largest numbers of births occurred in the third quarter (July to September), suggesting peak numbers of conceptions in the preceding winter months of November to February. Births were lowest in the fourth quarter (October to December) in six years, and the first quarter of the year (January to March) in the remaining four years. During the study period there was a mean difference in birth numbers between the lowest and highest quarters of 7.5% (range 5.8 to 8.9%). Therefore, on the basis of these results, and if no specific seasonality factors are operational, the maximum numbers of NTD and schizophrenia births would be expected in the spring and summer months, coincidental with the observed periods of maximum birth numbers.

## Neural tube defects

### Genetic factors

As indicated in the introduction, both genetic and environmental factors have been implicated in the aetiology of NTD. The evidence supporting a significant genetic contribution to NTD has several bases. Ethnic differentiation has been observed, with higher prevalence rates consistently reported for Whites than Blacks living under similar socioeconomic circumstances in the USA (Khoury et al., 1982; Windham & Edmonds, 1982). At the same time, studies in North America and Israel have shown that differences in NTD rates between national groups are modified by migration, although the variation in prevalence may persist into the second generation (Elwood & Elwood, 1980). With the exception of Northern Ireland where NTD rates are especially high (Little & Nevin, 1989), NTD births are more common in multiple as opposed to singleton pregnancies (Windham & Sever, 1982; Doyle et al., 1990; Imaizumi et al., 1991). Higher NTD rates have also been reported in the progeny of marriages between close biological relatives in Egypt (Stevenson et al., 1966), Algeria (Benallègue & Kedji, 1984), Tunisia (Khrouf et al., 1986), Iran (Naderi, 1979) and South India (Kulkarni et al., 1989). While suggestive of a major recessive gene contribution, in these regions the highest rates of consanguineous marriage are found in the poorest families (Bittles et al., 1991;

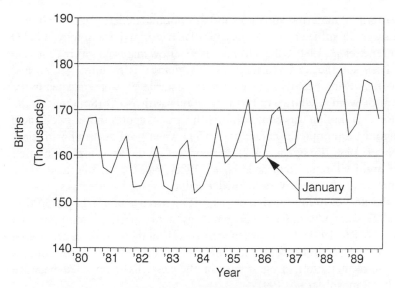

Fig. 9.1. Quarterly births in England and Wales, 1980–89.

Bittles, 1992), and so unless there has been adequate control for socioeconomic status the association between NTD prevalence and consanguinity must be interpreted with caution.

Despite the significant preponderance of males at conception and birth (Bittles *et al.*, 1993), globally there is a consistent excess of females diagnosed with NTD (Elwood & Elwood, 1980). The male:female ratio varies between localities and is greater with respect to anencephaly, ranging from 0.23 (Belfast) to 0.46 (Sydney), than in spina bifida with ratios of 0.34 (Manitoba) to 0.47 (Sydney). Finally, there is some evidence to suggest that in spina bifida, but not anencephaly, sibs are affected with similar types of lesion, i.e., that the disorder breeds true. However, there is no major support for the hypothesis that lesions above and below vertebral level 12, defined as neurulation and canalization defects respectively, have different genetic bases (Drainer *et al.*, 1991).

### Environmental factors

Although the list of proposed contributory environmental agents in NTD is extremely varied, many of the original suggestions implicating dietary agents, e.g., soft water (Lowe *et al.*, 1971), tea-drinking (Fedrick, 1974) and blighted potatoes (Renwick, 1972), have long been discounted. There is a well established negative association between NTD prevalence and socioeconomic status (Elwood & Elwood, 1980) that appears to be related to patterns of maternal nutrition, and sub-optimal maternal serum vitamin levels were identified as a possible causative factor in the aetiology of NTD (Laurence *et al.*, 1980; Schorah *et al.*, 1983). Low maternal serum folic acid has been specifically implicated, as this vitamin is required for normal growth and significantly reduced first trimester levels were detected in the serum of women who gave birth to infants with NTD (Smithells *et al.*, 1976). This finding was compatible with earlier reports that the folic acid antagonist aminopterin had caused NTD when used as an abortifacient (Meltzer, 1956; Emerson, 1962). Periconceptual dietary supplementation with multi-vitamins (Smithells *et al.*, 1980, 1983), and folic acid alone (Laurence *et al.*, 1981), were claimed to reduce the recurrence risks of NTD, and the efficacy of folic acid in preventing recurrence of a large proportion of NTD was subsequently confirmed by a prospective, multi-national survey (MRC, 1991). However, the precise role of the vitamin in promoting closure of the neural tube remains uncertain, and a number of periconceptual supplementation programmes in the USA have reported negative results (Simpson *et al.*, 1991).

In a small proportion of cases NTD may result from an iatrogenic effect. Retinoic acid, used for the treatment of severe cystic acne, has been identified as a potent teratogen and causative agent of NTD (Lammer *et al.*, 1985). Other drugs provisionally identified as causative agents of NTD include sodium valproate (Lindhout & Schmidt, 1986) and carbamazeprine (Rosa, 1991), both recommended in the control of epilepsy and, more controversially, clomiphene, which is administered for the induction of ovulation (Volsett, 1990).

### Birth seasonality in neural tube defects

Seasonality in the prevalence of central nervous system malformations was first suggested by McKeown & Record in 1951. To test the validity of their hypothesis, data on the months of birth of liveborn and stillborn infants with anencephaly and spina bifida in England and Wales were obtained from the OPCS for the decade 1980–89. For 1980 this information was available by quarter only and, because of changes in the OPCS recording procedure, the results for 1986 were incompatible with other years during the decade and so could not be included in the analysis. The numbers of NTD, livebirths and stillbirths combined, delivered per quarter between 1980–85 and 1987–89 are presented in Fig. 9.2. Of the 6586 NTD cases, 2152 were anencephalic (including 1056 stillbirths) and 4434 infants had spina bifida (including 846 stillbirths). There was a notable decline in NTD prevalence during the decade, from a total of 1642 cases in 1980 to 204 in 1989, reflecting the combined effects of a downward secular trend

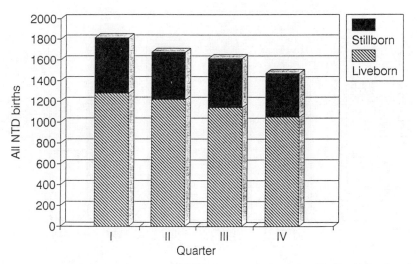

Fig. 9.2. Neural tube defects by quarter, England and Wales, 1980–85 and 1987–89.

(EUROCAT, 1991), improved antenatal detection and therapeutic abortion, and the gradual introduction of periconceptual vitamin supplementation. The changes were more pronounced for anencephaly, with decreases in the numbers of livebirths and stillbirths with the disorder of 90.1 and 92.7%; comparable figures for the reduction in cases of spina bifida were 82.1 and 95.1% respectively.

There were 23.2% more cases of NTD born during the first quarter of the study period (January to March) than in the fourth quarter (October to December). The difference in NTD numbers born by quarter was statistically significant for livebirths (chi$^2$ = 7.85 with 3 d.f., $p < 0.05$), and livebirths plus stillbirths (chi$^2$ = 11.02 with 3 d.f., $p < 0.025$), but not for stillbirths. Referring back to the data on total births in England and Wales during 1980–9, most deliveries occurred in the period July to September with fewest births between October and December (Fig. 9.3). The difference in numbers of births in each quarter was highly significant (chi$^2$ = 252.64 with 3 d.f., $p < 0.001$). Since the pattern of seasonality in NTD births did not coincide with the peak period of births in England and Wales as a whole, in fact the two maxima were six months apart, it cannot be dismissed merely as a simple statistical artifact.

To ascertain whether the pattern of NTD births could be ascribed to a particular winter month or months, the data from 1981–5 and 1987–9 were disaggregated and replotted, with no correction for the number of days per month. As shown in Fig. 9.4, high rates of NTD were reported in each month of the first quarter and into April. Although there is no obvious

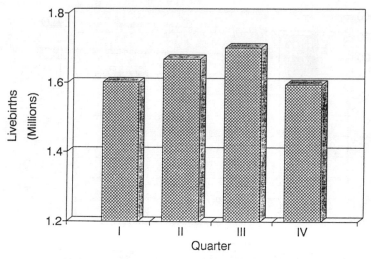

Fig. 9.3. Total livebirths, England and Wales, 1980–89.

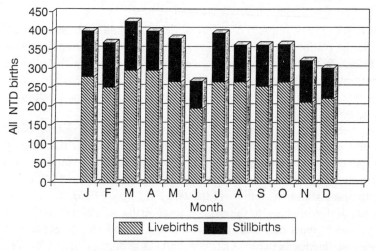

Fig. 9.4. Neural tube defects by month, England and Wales, 1981–85 and 1987–89.

explanation for the very low rate in June, other than random fluctuation due to the relatively small numbers analysed, the overall monthly trend from high to low prevalence throughout the year is apparent.

The demonstrated efficacy of folic acid dietary supplementation clearly implicates maternal folate status in the aetiology of NTD. Neural tube closure is completed within the first 28 days of pregnancy, and attempts to prevent the recurrence of NTD by vitamin supplementation after this time have been unsuccessful (Sheppard *et al.*, 1989). The peak timing of NTD births within the first quarter of the year points to an association with conceptions that have taken place in the second quarter of the preceding year. In northern temperate regions late spring was the time of year when food stocks were lowest, as it coincided with depletion of the winter stores and before fresh, vitamin-rich summer produce became available. It also was a time of intense agricultural activity, with planting of the new seasons's crops, which placed additional strains on the nutritional status of the population. However, maternal vitamin status appears not to be the sole determining factor since, in prospective studies into NTD recurrence, a small percentage of women conceived infants with spina bifida and anencephaly despite full periconceptual folic acid supplementation (MRC, 1991).

### Schizophrenia

The assessment of seasonality in schizophrenia presents a rather more difficult problem, owing to the adult onset of the disorder and the varying criteria employed in its clinical diagnosis (Kinney, 1991). Since the first report by Hare *et al.* (1972) the phenomenon has attracted considerable attention, due principally to the prevalence of the disorder worldwide. This has resulted in the accumulation of a large body of data, mainly collected in Western Europe and the USA.

#### Genetic factors

Contrary to the findings with NTD, studies conducted on the prevalence of schizophrenia in first cousin matings failed to detect a significant association with inbreeding (Ahmed, 1979; Saugstad & Ødegård, 1986), therefore the role of recessive genes in the disease appears to be of little significance. There are ethnic differences in prevalence, e.g., between Ashkenazi and Sephardi Jews in Israel (Dohrenwend *et al.*, 1992), and pedigree analysis in combination with extensive twin and adoption studies indicates a typical polygenic multifactorial mode of inheritance (Faraone & Tsuang, 1985), with concordance rates of 44.3% in monozygotic twins, 12.1% in dizygotic twins, and 7.3 and 9.4% in siblings and offspring (McGue *et al.*, 1985).

Following the introduction of molecular genetic techniques, the existence of a specific schizophrenia disease locus was initially suggested by the investigation in Canada of an uncle and nephew of Asian descent. Both were diagnosed schizophrenic, and on cytogenetic analysis they were shown to have an unbalanced translocation resulting in partial trisomy at chromosome 5q11.2 to 5q13.3 (Bassett *et al.*, 1988). Subsequent studies in Icelandic and English families resulted in the claim that schizophrenia was coinherited with chromosomal region 5q11–q13, thus indicating the presence of an allele predisposing to the disease (Sherrington *et al.*, 1988). To date this linkage has not been confirmed in Swedish, Scottish, Welsh or US kindreds (Kennedy *et al.*, 1988; St. Clair *et al.*, 1989; Detera-Wadleigh *et al.*, 1989; Aschauer *et al.*, 1990; McGuffin *et al.*, 1990; Crowe *et al.*, 1991), and it would appear that the original report either may have been a chance finding or a type 1 statistical error. Meanwhile, two other candidate regions have been proposed: a locus at 11q21–q22 associated with major mental illness including schizophrenia (St. Clair *et al.*, 1990) and, based on gender differences in clinical expression of the illness and familial risks, the pseudoanatomical region of the X and Y chromosomes (DeLisi & Crow, 1989).

### Environmental factors

The range of environmental variables that have been implicated as causative factors in schizophrenia is extremely wide. In the large majority of cases the hypothesis relies on epidemiological predictions, although contrary evidence frequently appears at least as strong. For example, a maternal age effect has been proposed (Dalen, 1988) and denied (Öhlund *et al.*, 1991), and seasonal temperature extremes also have been claimed (Hare & Moran, 1981) and refuted (Watson *et al.*, 1984). Obstetric complications of an undefined nature have been suggested (Bradbury & Miller, 1985), but the most convincing data relate to antenatal and/or perinatal exposure to viral infections, including cytomegalovirus (Torrey *et al.*, 1982), measles, varicella-zoster and polio (Torrey *et al.*, 1988), and influenza A2 (Mednick *et al.*, 1988; Kendell & Kemp, 1989). Several mechanisms have been proposed for the link between viruses and the symptoms of schizophrenia, including interference with fetal neural development (Mednick *et al.*, 1988), perinatal or childhood subclinical infections of the central nervous system (King *et al.*, 1985), and the triggering of immune dysfunction by infectious agents in the perinatal phase of development (Torrey *et al.*, 1988). Antenatal exposure to a virus(es) integrating at a site close to the cerebral dominance gene and interfering with its normal function also has been hypothesized (Crow, 1984), but the existence let alone the location of such a gene remains to be proven.

### Birth seasonality in schizophrenia

Extensive evidence for seasonality in the births of schizophrenics was collated by Bradbury & Miller (1985) and Boyd *et al.* (1986), who showed that in the northern hemisphere there was an identifiable excess of schizophrenia births in the winter months, mainly December to March, but

Table 9.1. *Surveys on birth seasonality in schizophrenia*

| Seasonality | North America | British Isles | Western Europe | Combined studies |
|---|---|---|---|---|
| Associated with births in winter (December to March) | 8 | 5 | 8 | 21 |
| Association with births in other months (April/May) | 0 | 2 | 0 | 2 |
| No association | 5 | 1 | 1 | 7 |
| Total numbers of surveys | 13 | 8 | 9 | 30 |
| Total numbers sampled | 32 917 | 42 608 | 62 233 | 142 758 |

See text for details of data sources cited.

occasionally with increased numbers born later in spring. Their findings, supplemented by data from England and Wales (Hare *et al.*, 1979), Ireland (O'Callaghan *et al.*, 1991), and France (D'Amato *et al.*, 1991) are summarized in Table 9.1. The total sample size comprised 142 758 cases of schizophrenia, with data collected on births between 1880 and 1979.

There was a positive association between a diagnosis of schizophrenia and birth during the months of December to March in 21 of the 30 surveys. In two studies, both conducted in the British Isles, an association between schizophrenia and season of birth was noted involving the months of April and May. Applying a Sign test to the total data set, the association between schizophrenia and winter births was significant at the $p < 0.05$ level. Interestingly, the results of comparable studies in Australia, New Zealand and South Africa have generally proved negative (Bradbury & Miller, 1985; Boyd *et al.*, 1986). These latter findings can be interpreted in terms of differing genetic predisposition, which seems improbable given the predominantly Western European origin of many populations in the southern hemisphere countries. Alternatively, if a virus(es) is indeed implicated in development of the schizophrenia phenotype, then the variant patterns of birth prevalence in the two hemispheres may stem from differences in the climatic range of the causative agent.

### Discussion

Although the genes involved in each of these disorders remain unknown, recognition of the associated environmental factors may help considerably in revealing their natural history. For NTD it has been appreciated for some time that, in areas of low prevalence, non-genetic factors appear to be of limited importance (Janerich & Piper, 1978; Sever, 1982), which would explain why periconceptual vitamin supplementation has had significantly greater effect in reducing recurrence rates in areas of high prevalence, such as Northern Ireland, than in the south-east of England (Seller & Nevin, 1984, 1990). One of the predictions associated with a polygenic multifactorial mode of inheritance is that in families where an individual of the usually less affected sex is conceived, there is the probability of a higher complement of risk genes (Carter, 1976). In keeping with this prediction, when NTD recurrences were observed despite maternal vitamin supplementation a majority of the cases were male (Seller & Nevin, 1984).

It appears clear that folic acid levels are of central importance in the prevention of NTD and a number of possible mechanisms for this association have been suggested. For example, since folic acid coenzymes are essential in purine and pyrimidine synthesis and hence nucleic biosynthesis, it has been proposed that a deficiency of folic acid at early stages of pregnancy may cause faulty cell division leading to NTD (Slattery & Janerich, 1991). However, the pre-existence of folic acid deficiency is far from certain as normal maternal folic acid levels have been reported during the second trimester in women carrying NTD fetuses, with no significant difference between blood levels of the vitamin in normal fetuses and those with NTD (Holzgreve *et al.*, 1991). Other hypotheses include defective folate metabolism (Yates *et al.*, 1987), possibly associated with reduced vitamin B12 levels (Gardiki-Kouidou & Seller, 1988), and the possibility that women who are at high risk of having a child with NTD are heterozygotes for homocystinuria, an inborn error of the trans-sulphuration pathway (Steegers-Theunissen *et al.*, 1991). More speculatively, the high-affinity folate-binding protein present in human amniotic fluid and thought to be derived from the placental folate receptor (Holm *et al.*, 1990) may prove to play a critical role in ensuring that appropriate levels of the vitamin are targetted to the embryo to permit closure of the neural tube in synchrony with the developmental timetable.

Why female embryos are at significantly higher risk of developing neural tube defects remains a mystery, especially since a large excess of male embryos are conceived (Bittles *et al.*, 1993). As noted above, empirical data show that the familial risk is greatest when the proband is male. Thus it could be argued that for females non-genetic factors generally may be more important in NTD, and so female conceptuses should benefit disproportionately from maternal dietary supplementation with folic acid. If this reasoning proves to be correct then, in those developed countries with periconceptual folate administration programmes, although fewer NTD cases will be conceived an increasing proportion will be vitamin non-responsive and male. In turn, the birth seasonality effect associated with NTD will become less obvious.

The situation with regard to seasonality in the births of schizophrenics is more complex because of difficulties resulting from the differential criteria used in diagnosis of the disease, coupled with disease heterogeneity due to probable multiple causality of both genetic and non-genetic origins. As with NTD, schizophrenia can be sub-divided into primarily genetic and non-genetic categories and the link between schizophrenia and seasonality is thought to be more common in non-familial, i.e., non-genetic, forms of the disease (Crow, 1984; Boyd *et al.*, 1986; Hsieh *et al.*, 1987). Despite evidence that contradicts this idea (Baron & Gruen, 1988), it has received strong support from studies in Ireland and France both of which propose that the seasonality effect is in fact restricted to non-familial forms of the disease (D'Amato *et al.*, 1991; O'Callaghan *et al.*, 1991).

Claims that the apparent birth seasonality of schizophrenia is spurious, due to age-effect distortions in prevalence data and errors in the design and interpretation of seasonal studies (Lewis, 1989), have provoked considerable controversy. The age-effect argument posits that, if the incidence of a disease increases with age, then since people born in the months of January to March are by definition older than those born later in the same year they should produce more new cases of the disease. However, this suggestion has been vigorously refuted by other workers who have counter-claimed misinterpretation both of their data and methods (Pulver *et al.*, 1990; Watson, 1990) and, as shown in Table 9.1, the weight of evidence does support a winter birth effect.

By comparison with NTD, there is as yet no convincing proof of an environmental agent that can be shown to be responsible for the symptoms of schizophrenia. The more extended season associated with the births of schizophrenics in the northern hemisphere, from December possibly through to April/May, and a lack of association in the southern hemisphere, makes the task of identification very difficult. The problem is exacerbated by uncertainty surrounding the timing of action of putative environmental agents. For example, influenza A2 has been implicated as a causative 'schizophrenogenic' agent acting in the second trimester of pregnancy (Mednick *et al.*, 1988), with the suggestion that maternal infection with the virus adversely affects development of the fetal brain, resulting in damage that predisposes the individual to schizophrenia decades later (Sham *et al.*, 1992). Whether the mechanisms of viral damage are direct or indirect, perhaps resulting from the body's response to viral attack, is open to speculation. Since reliance solely on epidemiological predictors may be seriously misleading, rigorous empirical testing of the various alternative hypotheses is essential if further progress is to be made in this line of investigation.

From a more general perspective, the results obtained in the studies on these two widely variant disease states suggest that birth seasonality could be a significant feature of other polygenic multifactorial disorders. Obviously, the greater the level of heterogeneity in the expression of a disorder the more difficult the phenomenon will be to detect. However, in terms of facilitating an improvement in our understanding of the natural history of common disease states, including ischaemic heart disease and rheumatoid disorders, such studies would clearly be merited.

## Acknowledgement

The award of a European Community ERASMUS Training Studentship to LS is acknowledged with gratitude.

## References

Ahmed, A.H. (1979). Consanguinity and schizophrenia in Sudan. *British Journal of Psychiatry*, **134**, 635–6.

Aschauer, H.N., Aschauer-Treiber, G., Isenberg, K.E., Todd, R.D. *et al.* (1990). No evidence for linkage between chromosome 5 markers and schizophrenia. *Human Heredity*, **40**, 109–15.

Baron, M. & Gruen, R. (1988). Risk factors in schizophrenia. Season of birth and family history. *British Journal of Psychiatry*, **152**, 460–5.

Bassett, A.S., McGillivray, B.C., Jones, B.D. & Pantzar, J.T. (1988). Partial trisomy chromosome 5 cosegregating with schizophrenia. *Lancet*, **i**, 799–800.

Benallègue, A. & Kedji, F. (1984). Consanguinité et santé publique. Étude algérienne. *Archives Francaises de Pédiatrie*, **41**, 435–40.

Bittles, A.H. (1992). Consanguinity: a major variable in studies on North African reproductive behavior, morbidity and mortality? In *Proceedings of the Demographic and Health Surveys World Conference*, ed. S. Moore, pp. 321–41. Columbia, MD: IRD/Macro.

Bittles, A.H., Mason, W.M., Greene, J. & Appaji Rao, N. (1991). Reproductive behavior and health in consanguineous marriages. *Science*, **252**, 789–94.

Bittles, A.H., Mason, W.M., Singarayer, D.N., Shreeniwas, S. *et al.* (1993). Determinants of the sex ratio in India: studies at national, state and local levels. In *Urban Ecology and Health in the Third World*, ed. L. Schell, M.T. Smith & A. Bilsborough, pp. 244–59. Cambridge: Cambridge University Press.

Boyd, J.H., Pulver, A.E. & Stewart, W. (1986). Season of birth: schizophrenia and bipolar disorder. *Schizophrenia Bulletin*, **12**, 173–86.

Brackenridge, C.J. (1980a). Bimodal month of birth distribution in cystic fibrosis. *American Journal of Medical Genetics*, **5**, 295–301.

Brackenridge, C.J. (1980b). Characteristics of births in each cycle of the bimodal monthly distribution in cystic fibrosis. *American Journal of Medical Genetics*, **5**, 303–7.

Bradbury, T.N. & Miller, G.A. (1985). Season of birth in schizophrenia: a review of evidence, methodology, and etiology. *Psychological Bulletin*, **98**, 569–94.

Carter, C.O. (1976). Genetics of common single malformations. *British Medical Bulletin*, **32**, 21–6.

Crow, T.J. (1984). A re-valuation of the viral hypothesis: is psychosis the result of retroviral integration at a site close to the cerebral dominance gene? *British Journal of Psychiatry*, **145**, 243–53.

Crowe, R.R., Black, D.W., Wesner, R. & Andreasen, N.C. (1991). Lack of linkage to chromosome 5q11–q13 markers in six schizophrenia pedigrees. *Archives of General Psychiatry*, **48**, 357–61.

Dalen, P. (1988). Schizophrenia, season of birth, and maternal age. *British Journal of Psychiatry*, **153**, 727–33.

D'Amato, T., Dalery, J., Rochet, T., Terra, J.L. *et al.* (1991). Saisons de naissance et psychiatrie. Étude rétrospective d'une population hospitalière. *L'Encéphale*, **17**, 67–71.

DeLisi, L.E. & Crow, T.J. (1989). Evidence for a sex chromosome locus for schizophrenia. *Schizophrenia Bulletin*, **15**, 431–40.

Detera-Wadleigh, S.D., Goldin, L.R., Sherrington, R., Encio, I. *et al.* (1989). Exclusion of linkage to 5q11–13 in families with schizophrenia and other psychiatric disorders. *Nature*, **340**, 391–3.

Dohrenwend, B.P., Levav, I., Shrout, P.E., Schwartz, S. *et al.* (1992). Socioeconomic status and psychiatric disorders: the causation–selection issue. *Science*, **255**, 946–52.

Doyle, P.E., Berel, V., Botting, B. & Wale, C.J. (1990). Congenital malformations in twins in England and Wales. *Journal of Epidemiology and Community Health*, **45**, 43–8.

Drainer, E., May, H.M. & Tolmie, J.L. (1991). Do familial neural tube defects breed true? *Journal of Medical Genetics*, **28**, 605–8.

Elwood, J.M. & Elwood, J.H. (1980). *Epidemiology of Anencephalus and Spina Bifida*. Oxford: Oxford University Press.

Emerson, D.G. (1962). Congenital malformations due to attempted abortion with aminopterin. *American Journal of Obstetrics and Gynecology*, **84**, 356–7.

EUROCAT Working Group (1991). Prevalence of neural tube effects in 20 regions of Europe and the impact of prenatal diagnosis, 1980–1986. *Journal of Epidemiology and Community Medicine*, **45**, 52–8.

Faraone, S.V. & Tsuang, M.T. (1985). Quantitative models of the genetic transmission of schizophrenia. *Psychology Bulletin*, **98**, 41–66.

Fedrick, J. (1974). Anencephalus and maternal tea drinking: evidence for a possible association. *Proceedings of the Royal Society of Medicine*, **67**, 6–10.

Gardiki-Kouidou, P. & Seller, M.J. (1988). Amniotic fluid folate, vitamin B12 and transcobalamins in neural tube defects. *Clinical Genetics*, **33**, 441–8.

Goad, W.B., Robinson, A. & Puck, T.T. (1976). Incidence of aneuploidy in a human population. *American Journal of Human Genetics*, **28**, 62–8.

Hare, E.H. & Moran, P. (1981). A relation between seasonal temperature and the birth rate of schizophrenic patients. *Acta Psychiatrica Scandinavica*, **63**, 396–405.

Hare, E.H., Bulusu, L. & Adelstein, A. (1979). Schizophrenia and season of birth. *Population Trends*, **17**, 9–11.

Hare, E.H., Price, J.S. & Slater, E. (1972). Schizophrenia and season of birth. *British Journal of Psychiatry*, **120**, 124–5.

Holm, J., Hansen, S.I. & Høier-Madsen, M. (1990). A high-affinity folate binding protein in human amniotic fluid. Radioligand binding characteristics, immunological properties and molecular size. *Bioscience Reports*, **10**, 79–85.

Holzgreve, W., Tercanli, S. & Pietrzik, K. (1991). Vitamins to prevent neural tube defects. *Lancet*, **338**, 640–1.

Hsieh, H.H., Khan, M.H., Atwal, S.S. & Cheng, S.C. (1987). Seasons of birth and subtypes of schizophrenia. *Acta Psychiatrica Scandinavica*, **75**, 373–6.

Imaizumi, Y., Yamamura, H., Nishikawa, M., Matsuoka, M. & Moriyama, I. (1991). The prevalence at birth of congenital malformations at a maternity hospital in Osaka city, 1949–1990. *Japanese Journal of Human Genetics*, **36**, 275–87.

Janerich, D.T. & Piper, J. (1978). Shifting genetic patterns in anencephaly and spina bifida. *Journal of Medical Genetics*, **15**, 101–5.

Kendell, R.E. & Kemp, I.W. (1989). Maternal influenza in the etiology of schizophrenia. *Archives of General Psychiatry*, **46**, 878–82.

Kennedy, J.L., Giuffra, L.A., Moises, H.W., Cavalli-Sforza, L.L. *et al.* (1988). Evidence against linkage of schizophrenia to markers on chromosome 5 in a northern Swedish pedigree. *Nature*, **336**, 167–70.

Khoury, M.J., Erickson, J.D. & James, L.M. (1982). Etiologic heterogeneity of neural tube defects: clues from epidemiology. *American Journal of Epidemiology*, **115**, 538–48.

Khrouf, N., Spang, R., Podgorna, T., Miled, S.B. *et al.* (1986). Malformations in 10,000 consecutive births in Tunis. *Acta Paediatrica Scandinavica*, **75**, 534–9.

King, D.J., Cooper, S.J., Earle, J.A., Martin, S.J. *et al.* (1985). A survey of serum antibodies to eight common viruses in psychiatric patients. *British Journal of Psychiatry*, **147**, 137–44.

Kinney, D.K. (1991). Schizophrenia and major affective disorders (manic-depressive illness). In *Principles and Practice of Medical Genetics*, 2nd edn., ed. A.E.H. Emery & D.L. Rimoin, pp. 457–72. Edinburgh: Churchill Livingstone.

Kulkarni, M.L., Mathew, M.A. & Reddy, V. (1989). The range of neural tube defects in southern India. *Archives of Diseases in Childhood*, **64**, 201–4.

Lammer, E.J., Chen, D.T., Hoar, R.M., Agnish, N.D. *et al.* (1985). Retinoic acid embryopathy. *New England Journal of Medicine*, **313**, 837–41.

Laurence, K.M., James, N., Miller, M. & Campbell, H. (1980). Increased risk of recurrence of pregnancies complicated by fetal neural tube defects in mothers receiving poor diets, and possible benefit of dietary counselling. *British Medical Journal*, **281**, 1592–4.

Laurence, K.M., James, N., Miller, M., Tennant, G.B. *et al.* (1981). Double-blind randomised controlled trial of folate treatment before conception to prevent recurrence of neural-tube defects. *British Medical Journal*, **282**, 1509–11.

Lewis, M.S. (1989). Age incidence and schizophrenia: Part I. The season of birth controversy. *Schizophrenia Bulletin*, **15**, 59–73.

Lindhout, D. & Schmidt, D. (1986). In-utero exposure to valproate and neural tube defects. *Lancet*, **i**, 1392–3.

Little, J. & Nevin, N.C. (1989). Congenital anomalies in twins in Northern Ireland. II: Neural tube defects. *Acta Genetica Medica Gemellologiae (Roma)*, **38**, 17–25.

Lowe, C.R., Roberts, C.L. & Lloyd, S. (1971). Malformations of the central nervous system and softness of local water supplies. *British Medical Journal*, **2**, 357–61.

McGue, M., Gottesman, I.I. & Rao, D.C. (1985). Resolving genetic models for the transmission of schizophrenia. *Genetic Epidemiology*, **2**, 99–110.

McGuffin, P., Sargeant, M., Hetti, G., Tidmarsh, S. *et al.* (1990). Exclusion of a schizophrenia susceptibility gene from the chromosome 5q11–q13 region: new data and a reanalysis of previous reports. *American Journal of Human Genetics*, **47**, 524–35.

McKeown, I. & Record, R.G. (1951). Seasonal variation in the frequency of anencephalus and spina bifida births in the United Kingdom. *Lancet*, **i**, 192–5.

Mednick, S.A., Machon, R.A., Huttunen, M.O. & Bonett, D. (1988). Adult schizophrenia following prenatal exposure to an influenza epidemic. *Archives of General Psychiatry*, **45**, 189–92.

Meltzer, H.J. (1956). Congenital anomalies due to attempted abortion with 4-aminopteroglutamic acid. *Journal of the American Medical Association*, **161**, 1253.

MRC Vitamin Study Research Group (1991). Prevention of neural tube defects: results of the Medical Research Council Vitamin Study. *Lancet*, **338**, 131–7.

Naderi, S. (1979). Congenital abnormalities in newborns of consanguineous and nonconsanguineous parents. *Obstetrics and Gynecology*, **53**, 195–9.

O'Callaghan, E., Gibson, T., Colohan, H.A., Walshe, D. *et al.* (1991). Season of birth in schizophrenia. Evidence for confinement of an excess of winter births to patients without a family history of mental disorder. *British Journal of Psychiatry*, **158**, 764–9.

Öhlund, L.S., Öhman, A., Öst, L.G., Lindström, L.H. *et al.* (1991). Electrodermal orienting response, maternal age, and seasons of birth in schizophrenia. *Psychiatry Research*, **36**, 223–32.

OPCS (1990). *Birth Statistics, Review of the Registrar General on Births and Patterns of Family Building in England and Wales, 1989*. London: Office of Population Censuses and Surveys.

Pulver, A.E., Moorman, C.C., Brown, C.H., McGrath, J.A. *et al.* (1990). Age-incidence artifacts do not account for the season-of-birth effect in schizophrenia. *Schizophrenia Bulletin*, **16**, 13–15.

Renwick, J.H. (1972). Hypothesis: anencephaly and spina bifida are usually preventable by avoidance of a specific but unidentified substance present in certain potato tubers. *British Journal of Preventive and Social Medicine*, **26**, 67–88.

Rosa, F.W. (1991). Spina bifida in infants of women treated with carbamazepine. *New England Journal of Medicine*, **324**, 674–7.

St. Clair, D., Blackwood, D., Muir, W., Baillie, D. *et al.* (1989). No linkage of chromosome 5q11–13 markers to schizophrenia and other psychiatric disorders. *Nature*, **339**, 305–9.

St. Clair, D., Blackwood, D., Muir, W., Carothers, A. *et al.* (1990). Association within a family of a balanced autosomal translocation with major mental illness. *Lancet*, **336**, 13–16.

Saugstad, L. & Ødegård, Ø. (1986). Inbreeding and schizophrenia. *Clinical Genetics*, **30**, 261–75.

Schorah, C.J., Wild, J., Hartley, R., Sheppard, S. *et al.* (1983). The effect of periconceptual supplementation on blood vitamin concentrations in women at recurrence risk for neural tube defect. *British Journal of Nutrition*, **49**, 203–11.

Seller, M.J. & Nevin, N.C. (1984). Periconceptual vitamin supplementation and the prevention of neural tube defects in South-East England and Northern Ireland. *Journal of Medical Genetics*, **21**, 325–30.

Seller, M.J. & Nevin, N.C. (1990). Vitamins during pregnancy and neural tube defects. *Journal of the American Medical Association*, **263**, 2749.

Sever, L.E. (1982). An epidemiologic study of neural tube defects in Los Angeles County. II. Etiologic factors in an area with low prevalence at birth. *Teratology*, **25**, 323–34.

Sham, P.C., O'Callaghan, E., Takei, N., Murray, G.K. *et al.* (1992). Schizophrenia following pre-natal exposure to influenza epidemics between 1939 and 1960. *British Journal of Psychiatry*, **160**, 461–6.

Sheppard, S., Nevin, N.C., Seller, M.J., Wild, J. *et al.* (1989). Neural tube defect recurrence after 'partial' vitamin supplementation. *Journal of Medical Genetics*, **26**, 326–9.

Sherrington, R., Bryjolffson, J., Petursson, H., Potter, M. *et al.* (1988). Localization of a susceptibility locus for schizophrenia on chromosome 5. *Nature*, **336**, 164–7.

Simpson, J.L., Mills, J.L., Rhoads, G.G., Cunningham, G.C. *et al.* (1991). Vitamins, folic acid and neural tube defects: comments on investigations in the United States. *Prenatal Diagnosis*, **11**, 641–8.

Slattery, M.L. & Janerich, D.T. (1991). The epidemiology of neural tube defects: a review of dietary intake and related factors as etiologic agents. *American Journal of Epidemiology*, **133**, 526–40.

Smithells, R.W., Sheppard, S. & Schorah, C.J. (1976). Vitamin deficiencies and neural tube defects. *Archives of Disease in Childhood*, **51**, 944–50.

Smithells, R.W., Sheppard, S. Schorah, C.J., Seller, M.J. *et al.* (1980). Possible prevention of neural-tube defects by periconceptual vitamin supplementation. *Lancet*, **i**, 339–40.

Smithells, R.W., Nevin, N.C., Seller, M.J., Sheppard, S. *et al.* (1983). Further experience of vitamin supplementation for prevention of neural tube defect recurrences. *Lancet*, **i**, 1027–31.

Steegers-Theunissen, R.P.M, Boers, G.H.J., Trijbels, F.J.M. & Eskes, T.K.A.B. (1991). Neural-tube defects and derangement of homocysteine metabolism. *New England Journal of Medicine*, **324**, 199–200.

Stevenson, A.C., Johnston, H.A., Stewart, M.I.P. & Golding, D.R. (1966). Congenital malformations: a report of a study of series of consecutive births in 24 centres. *Bulletin of the World Health Organization*, **34** (supplement), 1–125.

Torrey, E.F., Rawlings, R. & Waldman, I.N. (1988). Schizophrenic births and viral disease in two states. *Schizophrenia Research*, **1**, 73–7.

Torrey, E.F., Yolken, R.H. & Winfrey, C.J. (1982). Cytomegalovirus antibody in cerebrospinal fluid of schizophrenic patients detected by enzyme immunoassay. *Science*, **216**, 892–4.

Videbech, P. & Nielsen, J. (1984). Chromosome abnormalities and season of birth. *Human Genetics*, **65**, 221–31.

Volsett, S.E. (1990). Ovulation induction and neural tube defects. *Lancet*, **i**, 178.

Watson, C.G. (1990). Schizophrenic birth seasonality and the age-incidence artifact. *Schizophrenia Bulletin*, **16**, 5–12.

Watson, C.G., Kucala, T., Tilleskjor, C. & Jacobs, L. (1984). Schizophrenic birth seasonality in relation to the incidence of infectious diseases and temperature extremes. *Archives of General Psychiatry*, **41**, 85–90.

Windham, G.C. & Edmonds, L.D. (1982). Current trends in the incidence of neural tube defects. *Pediatrics*, **70**, 333–7.

Windham, G.C. & Sever, L.E. (1982). Neural tube defects among twin births. *American Journal of Human Genetics*, **34**, 988–98.

Yates, J.R.W., Ferguson-Smith, M.A., Shenkin, A., Guzman-Rodriguez, R. *et al.* (1987). Is disordered folate metabolism the basis for the genetic predisposition to neural tube defects? *Clinical Genetics*, **31**, 279–87.

# 10    *Environment, season and infection*

ANDREW TOMKINS

## Introduction

The seasonal burden of infections that affects many communities in developing countries has been recognised for many years by agricultural economists and health professionals. Disastrous epidemics of malaria were clearly described in different regions of India during the last century (Christophers, 1911). The extent of the population who became sick was so great that agricultural production was decreased. Moreover, so many were sick that transport and sale of food was limited. Food intakes decreased and famines developed. Similarly, populations afflicted by guinea worm showed seasonal patterns of infection of the large joints with profound effects upon the ability to farm. In both infections the health of entire populations was critically affected by the onset of rains that controlled rates of transmission of the parasites.

Health professionals in many communities have also noted striking differences in patterns of disease between seasons. These may be so noticeable that supplies and staffing arrangements are modified accordingly (Tomkins, 1981). The dramatic increase in cerebro-spinal meningitis during the dry seasons is sometimes so great that hospital wards become grossly overcrowded, necessitating the stockpiling of antibiotics and the erection of tents for temporary accommodation. The outbreaks of diarrhoeal disease, especially during the rainy season, require adequate supplies of oral rehydration salts and additional staff for care of adults and children with severe dehydrating diarrhoeal diseases such as cholera. Diarrhoeal treatment units are often full during the rainy season, numbers of admissions decreasing during the dry season. Syndromes of protein–energy malnutrition are often precipitated by measles and diarrhoeal disease; these account for the seasonal epidemics of syndromes such as marasmus and kwashiorkor, both conditions being more prevalent during the rainy season (Tomkins & Watson, 1989).

Several lines of research during the last few decades have concentrated on defining the mechanisms responsible for seasonal patterns of infection, and evaluating the impact of public health and clinical treatment strategies

123

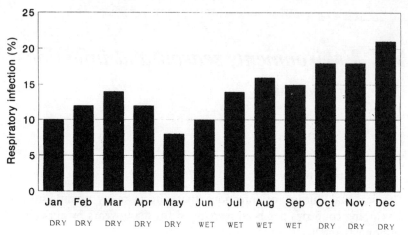

Fig. 10.1. Prevalence of respiratory infection among children aged 6/12 to 36/12 in the Gambia, W. Africa.

for prevention or management of seasonal epidemics. Some interventions, such as vaccines for meningococcal infection, have been remarkably successful (Greenwood & Wali, 1981). Others, such as attempts at malaria control by spraying of insecticides, have been rather disappointing (Bradley, 1991). The next few decades are likely to see increasing problems of crowding and inadequate sanitation and water supplies as population numbers increase and there are likely to be considerable increases in the intensity of seasonal patterns of infection. It will be particularly important to review the possible impact of different aspects of the environment (including temperature, rainfall, and toxins in the diet and atmosphere) and personal lifestyle on seasonal patterns of disease and to make a critical appraisal of interventions that might reduce the impact of seasonality on health, particularly among the poorest in society. It is, after all, the poorest who suffer the most from the vagaries of the environment (Chambers, 1981).

### Respiratory infections

Most respiratory infections peak during the dry season (Fig. 10.1). Several factors are likely to operate together in creating these peaks. A dry dusty climate alters the epithelium of the nose, throat, larynx and lungs. The additional effect of crowding, as a result of social gatherings and more closely packed sleeping patterns during the colder months is also important. Studies in Machakos, Kenya, clearly identified the seasonal epidemics of pertussis (Muller *et al.*, 1984) and measles (Leeuwenburg *et al.*, 1984). In recent years there has been considerable interest in the importance of host

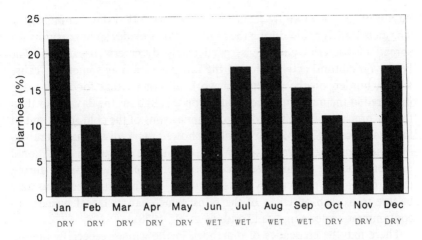

Fig. 10.2. Diarrhoea prevalence among children aged 6/12 to 36/12 in the Gambia, W. Africa.

nutrition in relation to respiratory infections such as measles. Indeed, the striking effect of Vitamin A supplementation on mortality from measles leads to considerations of whether improved Vitamin A status might prevent the development of other respiratory infections (Arthur *et al.*, 1992). Observational studies among young children in communities strongly suggest that Vitamin A deficient subjects are at risk of increased prevalence and complications of acute respiratory infections. Intervention studies indicate a causal relationship between Vitamin A deficiency and development of lower respiratory tract disease, particularly in infants with low birth weight. They also indicate a striking effect on morbidity and mortality of children with measles who have complications from pneumonia. Whether seasonal variations in Vitamin A status are sufficiently great to contribute to seasonal patterns of deaths from measles has not been demonstrated. At the present time it would seem that Vitamin A deficiency is more likely to influence whether a child dies, or develops complications, as opposed to whether a child develops an infection in the first place. Meningitis from meningococcus is characteristically severe during the dry months, epidemics usually ceasing at the onset of the rains (Blakeborough *et al.*, 1982).

### Diarrhoeal disease

Epidemics of diarrhoea are frequent during the rainy season (Fig. 10.2) as a result of the interaction of several factors. High temperatures and humidity facilitate the growth of diarrhoeal pathogens such that bacterial contamination of food is more frequent and more severe (Rowland *et al.*, 1978). The increased fly count during the rainy season provides an opportunity for spread of bacterial pathogens, particularly dysenteric organisms. The heavy agricultural activity during the rainy season may sometimes cause such a burden on the patterns of work of women that food has to be prepared in the morning for consumption throughout the day rather than freshly prepared at each meal time. Comparisons of the coliform count of food samples in Keneba, the Gambia, showed markedly greater numbers of bacteria during the rainy season than during the dry season. Additional problems include the increase in coliform count of water supplies during the rainy season, particularly in wells without protection. In part this may be due to a wash-off effect of the rains, which causes faecal bacteria to contaminate water supplies (Tomkins *et al.*, 1978).

There may be epidemics of diarrhoea in the winter, especially among infants and young children, as a result of rotavirus infection (Rowland *et al.*, 1980). This virus is transmitted by a variety of routes including aerosol and is the major pathogen responsible for epidemics of 'winter diarrhoea'.

### Malaria

The increase in prevalence of malaria parasites in the blood and clinical features of malaria during the rainy season is well described in many communities (Bradley, 1991). The dominant feature is the effect of climate and season on the behaviour pattern of the vector. *Anopheles gambiae* is a wet season breeder, responsible for much of the malaria in West Africa (Bray, 1981). *Anopheles melas* is also a wet season breeder and its breeding pattern is finely tuned by the salinity of the water in the mangrove swamps where it characteristically breeds (Giglioli, 1964). As the rains decrease the salinity, there is a critical level at which breeding may increase. This is followed by increasing rates of biting of human targets. It is interesting that the tidal patterns, determined by spring and neap tides, have an important effect on salinity in marshes, adding an additional component to the complexity of factors determining seasonal patterns of disease. Parasite prevalence and parasite density tend to increase in the blood of children during the wet months. The patterns of illness in which clinical diagnosis of malaria is made follow a similar course (Greenwood *et al.*, 1987).

### Filaria

Bancroftian filariasis involves infection with the nematode worm *Wuchereria bancrofti*. This parasite invades the lymphatics giving rise to a variety of clinical features such as swelling of the limbs, with considerable disability, in an economically important portion of the population. A variety of vectors are responsible, including *Anopheles gambiae* and *Anopheles funestus*. In Burkina Faso both species are involved and the number of people carrying microfilariae appears to be constant throughout the year. However, the biting activity of the mosquito increases during the rainy season so that the density of microfilariae per unit volume of blood almost doubles (Brengues, 1975). Interestingly, the clinical features do not have such a pronounced seasonal pattern. Lymphatic damage is probably accumulative.

### Onchocerciasis

The disabling condition 'river blindness' is caused by the nematode worm *Onchocerca volvulus*. This parasite is transmitted by the blackfly *Simulium damnosum* and invades the skin and eyes causing considerable disability and blindness. The fly requires high concentrations of oxygen for its own breeding pattern and the rush of water from hilly areas during the rains provides ideal breeding conditions in rivers, particularly those with waterfalls. The concentration of microfilariae in the skin reaches a peak during the rainy season in West Africa (Duke *et al.*, 1975).

### Kala-azar

The parasite *Leishmania* is transmitted by the bite of sandflies and produces a range of clinical syndromes, some so severe that there is fever, and enlargement of the liver and spleen. If untreated, mortality rates may be high. One of the vectors, *Phlebotomus martini*, has preference for the rainy season for its own breeding, and seasonal patterns of leishmaniasis are associated with increased biting rates (Southgate, 1964).

### Guinea-worm

The nematode *Dracunculus medinensis* lies under the skin. The worm causes painful lesions around the knees and ankles, so disabling that agricultural work is almost impossible during severe epidemics. Larvae are discharged from the skin into water supplies where they are taken up by the water flea *Cyclops*. These are then ingested by the next person and after burrowing through the stomach the female worm takes a year to mature and produces joint lesions which cause painful inflammation and discharge of larvae into water supplies, which infect the next person. Populations using step wells or surface water supplies are particularly at risk as these become infected very easily (Muller, 1979).

## Schistosomiasis

The three adult worms (*Schistosoma mansoni, haematobium* and *japonicum*) have characteristically different clinical features. *S. mansoni* and *S. japonicum* are excreted in the faeces whereas *S. haemotobium* is excreted in the urine. They then penetrate the intermediate host, the snail, and then discharge the infective form, the cercariae, into water; these can then penetrate the skin of people coming into contact with infected water supplies. Seasonal patterns of exposure to contaminated water are an important determinant of infection. In many communities, children tend to swim in ponds during the hot season. The decreasing amount of surface water in the dry season may lead to intensified water contact at times when cercariae are most likely to be present in the water. Acute symptoms may develop on exposure to the cercaria but the main clinical features of schistosomiasis develop as a result of an intensive load acquired over several years. Presentation of clinical illness, whether from haematuria or tiredness due to anaemia or abdominal pain due to enlargement of liver and spleen, do not have seasonal patterns. However, control programmes aiming to decrease the interaction between host and parasite need to consider seasonal patterns of infection.

## Tetanus

There may be seasonal patterns of tetanus in those communities where neonatal tetanus is the result of improper delivery practices such as the application of dirt and cow dung to the umbilical cord. The bacteria, and more importantly, the spores of the bacterium *Clostridium tetani* responsible for tetanus, favour a moist wet climate and they are destroyed by hot dry conditions. This may account for the seasonal peaks of tetanus during the rainy season in some communities.

## Skin infections

Lack of water supplies, particularly during the dry season, increases the tendency for spread of skin infections; in particular scabies is noted to have a marked seasonal pattern with increased risk during the dry season (Porter, 1977). This is serious in itself but has particularly important implications if the burrow into the skin left by the *sarcoptes* allows access by bacteria such as streptococci, which are responsible for the development of glomerulonephritis. Seasonal peaks of glomerulonephritis, secondary to scabies infection, are well recognised among children in West Africa.

Table 10.1. *Seasonality of infection in West Africa*

| Dry season | Wet season |
|---|---|
| Measles | Malaria |
| Meningitis | Typhoid |
| Pneumonia | Diarrhoea |
| Tetanus | Guinea-worm |
| Scabies | PEM/infection syndromes |
| Vitamin A deficiency | Anaemia |

## Malnutrition

The close interaction between malnutrition and infection requires careful consideration in assessing the genesis of seasonal patterns of infection (Waterlow *et al.*, 1992). Vitamin A deficiency is particularly likely to occur in the wet season as a result of decreased absorption when an individual has diarrhoea, and increased catabolism during systemic infection such as measles (Sommer, 1982). The seasonal availability of Vitamin A-containing foods such as mangoes could be an important factor in determining seasonality. Folate deficiency due to seasonal availability of vegetables and certain root crops such as yams has seasonal patterns. It could affect immune function sufficiently to increase susceptibility to infection but the relationship between folate deficiency and seasonality infection is largely unresearched (Tomkins, 1979).

## Classification of dry and wet season diseases

It should be possible to produce, for different geographical areas, a general pattern of diseases that are more prevalent in the dry or wet season. A typical example of that found in West Africa is shown in Table 10.1.

## Factors influencing seasonality of infection

During the next few decades several factors are likely to increase or decrease seasonal patterns of disease. Climatic change is likely to have striking effects. Already it has been noted that the transmission of schistosomiasis was reduced during the drought in the Sahel of the 1970s. Land pressure as a result of increasing population and the need to generate more food is likely to require intensive agricultural practices such as irrigation, which will provide increased opportunity for contact between people and cercariae. The prevalence of schistosomiasis is likely to increase. The increase in numbers of people living in towns and the resultant

Table 10.2. *Possible strategies for avoiding seasonal vector-borne infection*

| Bed nets | Sleeping away from sandflies |
| Sprays | Polystyrene balls against filariasis |
| Cattle penning | *Cyclops* control |

overcrowding is likely to increase the prevalence of respiratory infections.

Increasing urbanisation in poor countries may overcome existing sanitation systems. The increased prevalence of filaria infection in some urban communities during the last decade may well reflect the poor control of drainage and sanitation thereby facilitating the breeding of vectors.

While there are considerable possibilities for deterioration there are many for improvement. There have been notable developments in the range of drugs available for parasitic diseases. The development of praziquantel for schistosomiasis and albendazole for intestinal helminths now provides a remarkable opportunity for public health programmes even in those areas where environmental control has not yet become established. Vaccines are at present effective against pertussis, measles, tetanus and meningitis. Coverage rates within many developing countries have improved significantly in recent years. Experimental vaccines show some value for prevention of typhoid but there are no vaccines yet suitable for schistosomiasis, malaria, filaria or guinea-worm.

The global impact of human immunodeficiency virus infection on seasonality has yet to be calculated. There are considerable differences in the effect of HIV infection on different parasites, bacteria and viruses (Gilks *et al.*, 1992). Malaria infection, for instance, appears to be relatively little influenced by HIV infection whereas bacterial infections such as tuberculosis and bacterial causes of dysentery appear to show strikingly increased prevalence and severity.

New strategies for vector avoidance have been developed and field tested and show promising possibilities for use at a practical level in populations with few resources (Table 10.2). The reduction in prevalence of parasitaemia, clinical features and mortality from malaria as a result of using bed nets that have been impregnated with permethrin is striking (Procacci *et al.*, 1991). The use of personal sprays and repellants is effective but often uneconomic. Use of cattle penning to prevent access of vectors to humans and sleeping away from the customary habitats of sandflies such as termite hills are all useful strategies for reduction in transmission of leishmaniasis. Some remarkable results in terms of decreased transmission of filariasis have been achieved in Zanzibar as a result of the use of polystyrene balls that float on surface water sufficiently to prevent breeding of *Phlebotomus*

Table 10.3. *Factors influencing seasonality of infection – dysentery*

| Variable | Host/pathogen | Vector |
|---|---|---|
| Climate | + | + |
| Land pressure | − | + |
| Urbanisation | − | + |
| Chemotherapy | + | − |
| Vaccines | + | − |
| HIV | + | − |
| Vector avoidance | + | + |
| Nutrition | + | − |

Table 10.4. *Factors influencing seasonality of infection – malaria*

| Variable | Host/pathogen | Vector |
|---|---|---|
| Climate | + | + |
| Land pressure | − | + |
| Urbanisation | − | + |
| Chemotherapy | + | − |
| Vaccines | − | − |
| HIV | − | − |
| Vector avoidance | − | + |
| Nutrition | + | − |

*funestus,* the vector for filariasis. Unfortunately such public health/ engineering approaches do not seem so suitable for the control of *Anopheles gambiae.* A range of programmes against *Cyclops,* the vector of guinea worm, involving protection of water supplies and public health education, have been useful. Changes in behaviour patterns in relation to human/water contact are crucial in the prevention of schistosomiasis. Improving nutritional status is difficult but the demonstration that Vitamin A supplementation in Ghana is followed by a 38% decrease in hospital admissions gives an indication of the importance of micronutrients in infection (Arthur *et al.,* 1992).

**Future developments**

It is possible to produce some rather simplistic models in which the variables are examined for each infection. The three tables (10.3, 10.4, 10.5) show different patterns for measles, dysentery and malaria. It will be important for those developing policies for population, agriculture and health to consider the impact of change in each of the variables on the epidemiology of the infections. In certain instances a change in one variable, such as the more widespread availability of measles vaccine, could

Table 10.5. *Factors influencing seasonality of infection – measles*

| Variable | Host/pathogen | Vector |
|---|---|---|
| Climate | + | N/A |
| Land pressure | − | N/A |
| Urbanisation | + | N/A |
| Chemotherapy | + | N/A |
| Vaccines | + | N/A |
| HIV | − | N/A |
| Vector avoidance | − | N/A |
| Nutrition | + | N/A |

have profound effects on seasonality of severe protein–energy malnutrition and it is not unreasonable to hope that measles will be eliminated by the year 2000. Other variables such as urbanisation and density of housing are much more difficult to control. Any urban planner for instance will need to consider the increased patterns of infection that are likely to occur unless a range of environmental protection approaches are provided.

The development of new drugs could be a crucial factor in determining the seasonal patterns of infection. Effective drugs, readily available, could reduce the mortality from acute respiratory infection, typhoid and malaria. With the increasing difficulty that many governments face in being able to provide free health services it will be essential to explore ways of making safe, effective drugs available through a variety of well controlled commercial outlets. Early treatment is likely to have a striking effect on mortality.

Finally, it should be remembered that the decreasing prevalence of infections within industrialised countries over the last century did not only come about as a result of improved water supplies, hygiene, better housing and heating. There were striking improvements in nutrition and these may be important in reducing the prevalence of the seasonally determined infections. At the very least, approaches for improved household food security will be necessary if morbidity and mortality from infections such as diarrhoea and pneumonia are to be controlled. In previous years it has been customary to concentrate on intakes of energy and protein. It may well be that nutritionists need a broader view with greater emphasis on micronutrients in the future.

### References

Arthur, P., Kirkwood, B., Ross, D., Morris, S., Gyapong, J., Tomkins, A.M. & Hutton, A. (1992). Impact of vitamin A supplementation on childhood morbidity in northern Ghana. *Lancet*, **339**, 361–2.

Blakeborough, I.S., Greenwood, B.M., Whittle, H.C., Bradley, A.K. & Gilles, H.M. (1982). The epidemiology of Neisseria meningitis and Neisseria lactamica in a northern Nigerian community. *Journal of Infectious Diseases*, **146**, 626–37.

Bradley, D.J. (1991). Malaria. In *Disease and Mortality in Sub-Saharan Africa*, ed. R.G. Feachem & D.T. Jamison, pp. 190–202. Oxford University Press/The World Bank, Washington.

Bray, R.S. (1981). Insect-borne diseases: malaria. In *Seasonal Dimensions to Rural Poverty*, ed. R. Chambers, R. Longhurst & A. Pacey, pp. 116–31. London: Frances Pinter.

Brengues, J. (1975). La filaroise de Bancroft en Afrique de l'ouest. Paris: *Memoires*, ORSTOM, No. 79.

Chambers, R. (1981). In *Seasonal Dimensions to Rural Poverty*, ed. R. Chambers, R. Longhurst & A. Pacey, pp. 1–8. London: Frances Pinter.

Christophers, S.R. (1911). Malaria in the Punjab. *Scientific Memoirs by Officers of the Medical and Sanitary Departments of the Government of India*, New Series No. 46, Calcutta.

Duke, B.O.L., Anderson, J. & Fuglsang, H. (1975). Onchocera volvulus transmission potentials and associated patterns of Onchocerciasis at four Cameroon villages. *Tropenmedizin und Parasitologie*, **26**, 143–54.

Giglioli, M.E.C. (1964). Tides, salinity and the breeding of Anopheles melas in the Gambia. *Riv. Malario*, **43**, 245–62.

Gilks, C.F., Otieno, L.S., Brindle, R.J., Newnham, R.S., Luce, G.N., Were, J.B.O., Simani, P.M., Ghatt, S.M., Okelo, G.B.A., Waiyaki, P.G. & Warrell, D.A. (1992). The presentation and outcome of HIV-related disease in Nairobi. *Quarterly Journal of Medicine*, New Series, **82**(297), 25–32.

Greenwood, B.M. & Wali, S.S. (1981). Control of meningococcal meningitis in the African meningitis belt by selective vaccination. *Lancet*, **1**, 729–32.

Greenwood, B.M., Greenwood, A.M., Bradley, A.K., Byass, P., Jammeh, K., Marsh, K., Tulloch, S., Oldfield, F.S.J. & Hayes, P. (1987). Mortality and morbidity from malaria among children in a rural area of the Gambia, West Africa. *Transactions of the Royal Society of Tropical Medicine and Hygiene*, **81**, 478–86.

Leeuwenburg, J., Muller, A.S., Voohoeve, A.M., Gemert, W. & Kok, P. (1984). The epidemiology of measles. In *Maternal and Child Health in Rural Kenya: An Epidemiological Study*, ed. J.K. van Ginneken & A.S. Muller, pp. 1–225. London: Croom Helm.

Muller, A.S., Leeuwenburg, J. & Voorhoeve, A.M. (1984). Pertussis in a rural area of Kenya: epidemiology and results of a vaccine trial. *Bulletin of the World Health Organization*, **62**, 899–908.

Muller, R. (1979). Guinea worm disease: epidemiology, control and treatment. *Bulletin of the World Health Organization*, **57**, 683–9.

Porter, M.J. (1977). An epidemiological approach to skin diseases in the tropics. *Tropical Doctor*, **7**, 59–66.

Procacci, P.G., Lamizana, L., Kumlien, S., Habluetzel, A. and Rotgliano, G. (1991). Permethrin-impregnated curtains in malaria control. *Transactions of the Royal Society of Tropical Medicine and Hygiene*, **85**(2), 181.

Rowland, M.G.M., Barrell, R.A.E. & Whitehead, R.G. (1978). Bacterial contamination in traditional Gambian weaning foods. *Lancet*, **1**, 136–8.

Rowland, M.G.M., Leung, T.S.M. & Marshall, W.C. (1980). Rotavirus infection in young Gambian village children. *Transactions of the Royal Society of Tropical Medicine and Hygiene*, **74**, 663–5.

Sommer, A. (1982). *Nutritional Blindness*. Oxford: Oxford University Press.

Southgate, B.A. (1964). Studies in the epidemiology of East African leishmaniasis. *Transactions of the Royal Society of Tropical Medicine and Hygiene*, **58**, 377–90.

Tomkins, A.M. (1979). Folate malnutrition in tropical diarrhoea. *Transactions of the Royal Society of Tropical Medicine and Hygiene*, **73**, 498–502.

Tomkins, A.M. (1981). Seasonal health problems in the Zaria region. In *Seasonal Dimensions to Rural Poverty*, ed. R. Chambers, R. Longhurst & A. Pacey, pp. 177–81. London: Frances Pinter.

Tomkins, A.M. & Watson, F. (1989). A review prepared for ACC/SCN State of the Art Series Nutrition Policy Discussion, Paper No. 5, 1–135.

Tomkins, A.M., Drasar, B.S., Bradley, A.K. & Williamson, W.A. (1978). *Transactions of the Royal Society of Tropical Medicine and Hygiene*, **72**, 239–43.

Waterlow, J.C., Tomkins, A.M. & Grantham-McGregor, S.M. (1992). *Protein–Energy Malnutrition*. London, Melbourne, Aukland: Edward Arnold.

# 11  *Seasonal mortality in the elderly*

K.J. COLLINS

## Introduction

The United Nations, in their *Demographic Year Books*, publish, approximately every six years, details of monthly mortality statistics (deaths from all causes) gathered from many countries throughout the world. For many of these countries monthly mortality figures are strongly correlated with season. Seasonal fluctuations in mortality can be attributed to many factors depending on the region, for example to seasonal availability of food, to climatic phenomena such as monsoons, and to seasonality in the occurrence of disease epidemics. Climatic temperature itself is a primary component, with highest death rates coinciding with seasonal temperature extremes of heat in low latitudes and cold in high latitudes. Seasonal swings in morbidity and mortality are not confined simply to regions with extremes of temperature. Many temperate countries, including the British Isles, exhibit strong seasonal mortality trends (Fig. 11.1).

It is of considerable concern that there is an excess winter mortality in the British Isles that is of a magnitude much higher than in many other comparable countries, particularly Scandinavia and neighbouring parts of Western Europe. Many of these European countries experience much more severe cold winters than in Britain. Certain 'at-risk' groups such as infants, the sick and disabled, and the elderly, are the most vulnerable to the effects of seasonal temperature changes. In Great Britain the high excess winter mortality, especially in the elderly, is well recognised, but the aetiology of this phenomenon is not fully understood.

## Early studies on seasonal mortality

A brief mention of the influence of season on mortality was made in the *Second Annual Report of the Registrar General*, representing England in 1840. A more extensive discussion of the effect of low temperature on mortality in London (1838–41) appeared in the subsequent report of 1841. It was observed that the numbers of deaths rose when the temperature at night fell below freezing, and rose more markedly when mean day and night

135

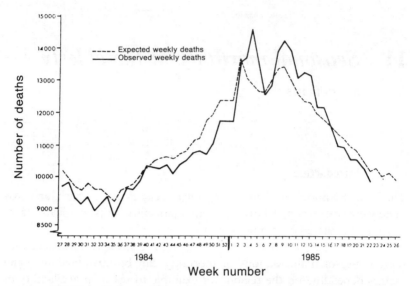

Fig. 11.1. Observed and expected weekly death registrations, excluding deaths of infants under one year old, for England and Wales, 1984–85 (from Alderson, 1985).

temperatures were at freezing point or below. The increase in mortality with cold weather occurred immediately, but the effects appeared to continue over a 30 to 40 day period afterwards. A later *Registrar General's Report* (1874) showed that the increase in mortality accompanying cold weather was much greater in the elderly, for age groups of age 40 and over, the excess deaths doubled for every extra nine years in age. It was suggested that exercise, nutritious food, artificial heat, appropriate clothing, and a respirator to retain heat from the respiratory tract might help to combat the effects of cold on elderly people.

Early epidemiological studies (e.g. Guy, 1881) established that cold winters increased mortality, particularly from respiratory disease in elderly people. Comparing Britain and the USA, it was found that mortality rates in England were higher in winter but lower in summer than in the USA, though the summer rates were comparable in England and in the states of New England (Lewis-Fanning, 1940). Mortality from cardiovascular diseases was also found to increase in the winter, and to be related to temperature, air pollution and rainfall (Rose, 1966). During the period 1950–62 the mid-winter peak in cardiovascular disease mortality varied from 20% to 70% above the mid-summer trough. The suggestion was put forward that changes in ambient temperature were responsible for most of the short-term fluctuations in deaths from ischaemic heart disease. Regressions of temperature against mortality for England and Wales, Australia, Canada and Denmark showed steeper gradients for diseases of

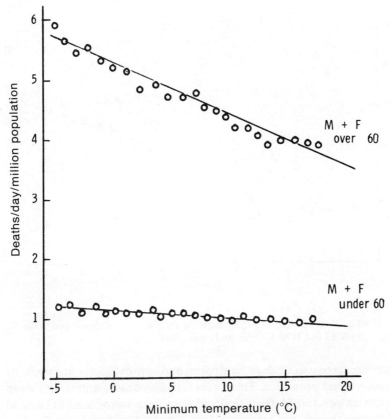

Fig. 11.2. Daily deaths from myocardial infarction for younger and older age groups in England and Wales 1970–71 (from Bull & Morton, 1975).

the respiratory system than for cardiovascular diseases (Anderson & Le Riche, 1970). It was pointed out that there were seasonal fluctuations in the incidence or severity of respiratory infections, which could have an influence on deaths from ischaemic heart disease. However, because of the wide differences in monthly mean temperature in the four countries it was doubted whether temperature variation alone accounted for the results observed.

Bull (1973) demonstrated that the association between myocardial infarction or cerebrovascular deaths and temperature in England and Wales was not a reflection of the association with respiratory or infectious diseases. Bull & Morton (1978) also emphasised the strong relationship between environmental temperatures and death rates for people older than 60 years (Fig. 11.2). The lengths of time between the onset of a cold spell and the increase in mortality was one to two days for myocardial

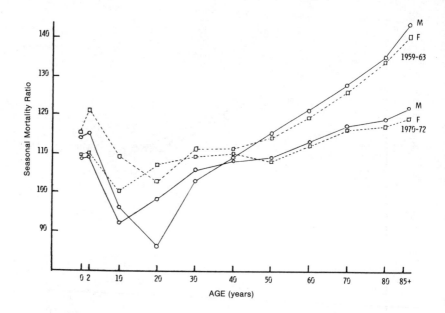

Fig. 11.3. Seasonal Mortality Ratio by age and sex in England and Wales for 1959–63 and 1970–72 (from McDowell, 1981).

infarction, three to four days for strokes, and approximately one week for pneumonia and bronchitis. There were significant associations with mean ambient temperature for the majority of the main causes of death in England and Wales, though not with neoplasms. Others have examined the time sequence and confirm a two-day time lag between a fall in temperature and an increase in both cardiac and stroke mortality (Rogot & Padgett, 1976).

### Indices of seasonal mortality

A procedure used to assess the extent of the seasonal effect on mortality is the seasonal mortality ratio (SMR). This can be expressed as the death rate for the coldest winter quarter (in Great Britain, the January to March quarter) divided by the average rate for the same calendar year taken as 100 (McDowell, 1981). The SMR was found to increase from an average of 110 for the period 1841–50 to 128 for the period 1921–30 in England and Wales. One reason for this increase over the 90-year period appears to be the progressive elimination of specific causes of summer mortality due to poor hygiene linked to enteric disease. However, since 1930, the SMR has been falling, with an average of 114 for the period 1979–88. A ratio of 114 implies some 20 000 extra deaths for the quarter. The possible effects of a reduction in air pollution and increasing use of central heating were considered by

Table 11.1. *Some international comparisons of the mean excess winter death index for the period 1976–84*

| | | | |
|---|---|---|---|
| England and Wales | 21 | Austria | 11 |
| Scotland | 20 | Bulgaria | 19 |
| Northern Ireland | 18 | East Germany | 10 |
| Irish Republic | 24 | Czechoslovakia | 9 |
| Denmark | 9 | Canada | 7 |
| Finland | 8 | USA | 9 |
| Norway | 9 | Israel | 26 |
| Sweden | 10 | Japan | 17 |
| Belgium | 11 | Australia | 20 |
| France | 11 | New Zealand | 25 |
| West Germany | 8 | | |
| Italy | 19 | | |
| Netherlands | 9 | | |
| Portugal | 26 | | |
| Spain | 19 | | |
| Switzerland | 10 | | |

From Curwen (1990/1).

McDowell to be involved in the lowering of winter excess mortality, though causality was found difficult to prove. It seemed, however, that fewer severe winters or less frequent winter influenza epidemics were not responsible. The SMR was also analysed in terms of sex differences and it was found to be slightly higher for males up to the fourth decade, after which the sex difference disappeared; there then followed a progressive rise for both sexes into old age (Fig. 11.3).

Another expression is the excess winter death index (EWDI) defined as the percentage excess of deaths in the four winter months of highest mortality in England and Wales (December–March) compared with the average of the numbers in the preceding and following four-monthly periods (Curwen, 1990/1). International mortality statistics highlight the fact that the EWDI is considerably higher in Great Britain than in many comparable countries (Table 11.1). In the 1960s and 1970s Britain had the highest EWDI for Europe and North America (Collins, 1987). This trend now appears to be declining though there remains a large discrepancy between Great Britain and North America and Scandinavia. In some countries, a low EWDI may be due to excess summer mortality. Table 11.1 illustrates the fact that those countries with mild winters (Israel, Portugal, New Zealand) show greater EWDIs than those with cold winters (Canada, Finland, Germany). The recent decline in the EWDI in Great Britain and other temperate countries has not until now been matched by a similar reduction in winter mortality in New Zealand, which suggests that useful epidemiological comparisons between the two countries would be of particular interest.

Table 11.2. *Excess winter death index by age and sex for England and Wales*

| Age (years) | Seven winters 1968–78 | | Eight winters 1977–85 | | One winter* 1984–85 | |
|---|---|---|---|---|---|---|
| | M | F | M | F | M | F |
| 45–64 | 19 | 16 | 13 | 11 | 15 | 15 |
| 65–74 | 25 | 25 | 18 | 18 | 22 | 23 |
| 75+ | 32 | 30 | 27 | 27 | 35 | 35 |

*The coldest winter since 1978–9.
From Curwen (1990/1).

### Excess winter deaths in the elderly

Regression analyses (OPCS, 1987; Curwen & Devis, 1988) have established that about 88% of the variation in numbers of excess winter deaths in England and Wales could be accounted for by three variables: mean national winter temperature, influenza epidemics, and a secular trend. Time was included in the analyses to account for the secular trend representing a reduction of excess winter deaths by about 500 per year, *ceteris paribus*. The secular trend, however, takes no account of ageing of the population. The proportion of excess winter deaths increases with age, currently of the order of 150 excess deaths per year due to the ageing effect alone (Curwen, 1990/1) (Table 11.2).

One interpretation of the regression model is that for every degree centigrade change in the average temperature in the winter there is an increase or decrease in the number of winter deaths by about 8000 (Alderson, 1985). Changes in ambient temperature are linked to other environmental factors such as humidity; these may be important but they are usually not specified in the national statistics. Each registered influenza death is associated with between three and four excess winter deaths, for example in the winter of 1975–6 when there was a major influenza epidemic, the total number of excess winter deaths was 58 000. Of these, 22 000 were *associated* with influenza; this included 6000 deaths registered as being *caused* by influenza. It has been calculated (Curwen, 1990/1) that during the period 1976–88 influenza was responsible for just under a quarter of all excess winter deaths.

Since the cold winter of 1963, attention has been focused on the issue of deaths from hypothermia as a special aspect of seasonal mortality in the elderly (Royal College of Physicians of London, 1966). There is strong evidence to show that although thermoregulation does not usually fail in the elderly, the physiological potential to respond to cold may diminish (Collins & Exton-Smith, 1983) and the aged are therefore more at risk in cold winters. Hospital surveys find that most elderly people admitted to hospital with deep body temperatures below 35 °C are suffering from secondary hypothermia that is associated with other intercurrent medical conditions (Collins, 1983). National statistics show that in recent years only a relatively small number of elderly people are diagnosed as having hypothermia. An EWDI of 14 implies some 20 000 extra deaths in the winter whereas there have never been more than about 700 deaths per year registered with a mention of hypothermia (Collins, 1983). Hypothermia deaths may account for about 1% of the excess winter mortality.

### Social factors

For elderly males aged 65–74 years, excess winter mortality increases significantly from Social Class I to Social Class V; for example, based on mortality statistics around the 1971 census (OPCS, 1987), EWDIs rose consistently from 15 (Social Class I) to 30 (Social Class V). Socio-economic demographic mortality differentials were examined in a longitudinal study based on a 1% sample of the whole population from the 1971 census followed up until death (Fox & Goldblatt, 1982). The linking of personal data recorded at the census and again at death makes possible the use of various social markers that are often more sensitive to socio-economic status than are those based simply on occupation. Social markers can include 1. housing tenure, for example owner-occupied compared to rented accommodation, and 2. access to a car at home. Curwen (1990/1) observed that for deaths in these groups from 1971 to 1985 in England and Wales, the EWDIs for owner-occupied householders with car access was 18 (for males), and 18 (for females) and for those in rented accommodation without car access, 22 (for males), and 25 (for females). There is evidence therefore to support the link between social class and the EWDI.

### Temperature of dwellings

A survey of world mortality statistics by Sakamoto-Momiyama (1977) indicated that a principal cause of the general reduction in seasonal swings during this century was the increasing use of central heating in dwellings. McDowell (1981) considered that the factors of improved home heating and reduced air pollution might have contributed to the fall in SMRs in the last two decades in England and Wales. It is difficult to find convincing evidence of a direct nature to prove causality, but it is difficult to ignore, for example, the strong negative relationship between winter mortality and average indoor temperatures in different European countries (Boardman, 1993).

142    K.J. Collins

Table 11.3. *Comparison of the coefficient of variation in seasonal mortality with the percentage of households with central heating in different regions of England and Wales in 1982*

| Region | Percentage of households with central heating | Seasonal variation in mortality |
|---|---|---|
| North | 63.4 | 0.519 |
| Yorks. and Humberside | 56.4 | 0.581 |
| East Midlands | 65.3 | 0.637 |
| East Anglia | 66.5 | 0.537 |
| South East | 63.2 | 0.509 |
| South West | 58.2 | 0.462 |
| West Midlands | 56.7 | 0.599 |
| North West | 54.6 | 0.610 |
| Wales | 57.8 | 0.481 |
| England and Wales | 60.2 | 0.508 |

From Alderson (1985).

Some studies (e.g. Keatinge, 1986) claim that there is no difference in winter mortality between elderly people living in centrally heated accommodation and those without. In another study in eight regions in England and Wales, Alderson (1985) observed no differences in winter mortality and the percentage of households with central heating for those over 75 years of age (Table 11.3). There are obvious pitfalls in associating data on seasonal mortality and indoor heating. For example, it is erroneous to assume that the possession of central heating automatically confers warm homes, for affordable warmth may not be available, especially for some elderly or those suffering fuel poverty, despite the presence of central heating installations (Boardman, 1986).

Low indoor temperatures can be harmful to the elderly, and in the winter in Great Britain temperatures in some dwellings may fall below 10 °C. It is recommended that 16 °C is a minimum indoor temperature in order to maintain health for the average person, with a 2–3 °C warmer minimum temperature for sedentary elderly (Collins, 1986). The contrast between high EWDI in Great Britain and other Northern European countries that have little seasonal mortality variation strongly suggests an effect of indoor heating. In Finland, for example, a substantial fall in seasonal mortality during the last 50 years has corresponded with greatly improved housing conditions, including heating (Nayha, 1984).

Central heating installations have increased in households in Great Britain from 13% to 66% in the period 1964 to 1984, and with them the EWDI has declined. Excess mortality from respiratory disease has decreased by 69% during this period, even when adjustments are made for

Table 11.4. *Mean excess winter deaths (EWDI) by age, sex and principal cause of death in England and Wales, 1976–81*

| Cause | | Mean EWDI by age | | | Mean annual no. of EWD |
|---|---|---|---|---|---|
| | | 45–64 | 65–74 | 75+ | 45+ |
| Ischaemic heart disease | M | 16 | 20 | 27 | 6216 |
| | F | 18 | 20 | 25 | 4696 |
| Cerebrovascular disease | M | 16 | 20 | 30 | 2030 |
| | F | 16 | 19 | 27 | 3287 |
| Other circulatory disease | M | 20 | 24 | 34 | 1690 |
| | F | 22 | 28 | 31 | 3341 |
| Respiratory disease | M | 53 | 53 | 53 | 6282 |
| | F | 54 | 57 | 58 | 6596 |
| Accident and violence | M | 8 | 15 | 36 | 316 |
| | F | 12 | 20 | 38 | 541 |
| All other causes | M | 3 | 5 | 11 | 1820 |
| | F | 3 | 5 | 9 | 1811 |

From Curwen (1990/1).

the varying coldness of winters (Keatinge *et al.*, 1990). Some of the improvement may be due to a decline in influenza epidemics for the period, reduced cigarette smoking, and the availability of better antibiotics. By contrast, excess winter mortality from coronary and cerebrovascular disease, though rising during some infectious winter epidemics early in the 1964–84 period, has not fallen significantly as home heating has improved.

The evidence available thus far suggests that changes in respiratory mortality in the elderly in winter are most closely associated with influenza epidemics and the effects of home heating. The immediate influence of outdoor cold may be more important than indoor temperature in increasing coronary and cerebrovascular deaths.

### Aetiological factors

Table 11.4 lists six broad categories of excess winter deaths in England and Wales (Curwen, 1990/1) showing two main causes: cardiovascular disease and respiratory disease. Cardiovascular deaths are mostly due to ischaemic heart disease (causing myocardial infarction), cerebrovascular disease (causing strokes) and 'other' circulatory causes. There is a strong age gradient for each type (Table 11.4). For the period 1976–81, 55% of all excess winter deaths for those over 45 years of age (and 52% of non-winter deaths) fall into the cardiovascular category. Cardiovascular disease is thus the most frequent cause of death at all times of the year. By contrast, the age gradient for respiratory disease is about 50% greater in the winter than during the rest of the year, and 33% of excess winter deaths (and 14% of non-winter deaths) are caused by respiratory disease. Accidents and violence (some associated with hypothermia) account for less than 2% of excess deaths over the age of 45 years, and there is a noticeable age gradient.

One of the underlying causes of excess cardiovascular mortality in the winter, it is suggested, is coronary and cerebral thrombosis as the result of increased blood viscosity, haematocrit and platelets in cold weather. The change in blood viscosity with cold exposure has been demonstrated in many laboratory studies (e.g. Burton & Edholm, 1955) and more recently with mild skin surface cooling (Keatinge *et al.*, 1984). Though these investigations have been carried out on young adults, it is postulated that in elderly people with more advanced atherosclerosis in blood vessels there is a greater likelihood of platelet deposition and thrombus formation. Plasma fibrinogen, a major contributor to blood viscosity and thrombus formation, was observed to be 23% higher in the winter than in summer in elderly people living in their own homes or sheltered dwellings in Northern Ireland (Stout & Crawford, 1991). There were significant negative associations between fibrinogen levels in blood and the indoor and outdoor temperatures. This is an important consideration in the aetiology of winter excess deaths, for prospective studies have shown that fibrinogen levels in the blood can predict the development of cardiovascular disease (Kannel *et al.*, 1987). Equally it is important to be cautious in interpretation of these results. Fibrinogen is also known to be an acute phase protein which is likely to increase in concentration in blood in the presence of more frequent winter infections.

Another factor that may play a part in causing a higher cardiovascular mortality in the elderly is the change in responsiveness of cardiorespiratory reflexes with age. In cold conditions, arterial blood pressure is found to increase more markedly in older people as the result of diminished baroreflex control (Collins *et al.*, 1985). Elderly people exposed to a cold air stream on the face when outdoors may be at greater risk than young adults. Facial cooling in air produces a reaction similar to the 'diving response' experienced during immersion of the face in cold water. The reflexes invoked – apnoea, bradycardia and peripheral vasoconstriction – are normally protective but may pose a circulatory threat when they become exaggerated or uncontrolled. It has been shown that there is a significantly greater rise in arterial blood pressure and a reduced slowing of the heart in old people compared to young when the face is cooled in an air stream at 3 °C (Collins *et al.*, 1989). In both elderly and young, these effects are enhanced by whole body cooling and diminished by body warming. There are numerous accounts of the effects of acute exposure to cold combined with unaccustomed physical exertion, such as shovelling snow, which have precipitated heart attacks or strokes, especially in the older, less-fit group (Rogot & Padgett, 1976).

The poor prognosis that often accompanies the long period of recovery following lower limb fracture in old people, often with the development of pneumonia, indicates the serious nature of accidents causing falls in the elderly. There is an increased incidence of fractures of the femoral neck in old people during the winter, with an increase of 80% in cases occurring outdoors and an increase of 350% indoors. Fractures are observed to increase in frequency within days of a decrease in environmental temperature (Allison & Bastow, 1983). It is possible that increases in falls may occur as the result of the effects of cold on sensory perception (intrinsically diminished in many elderly, Collins & Exton-Smith, 1983) and on co-ordination and muscle function. Under-nutrition also strongly affects morbidity and mortality in old people. There is a higher incidence of mid-winter falls indoors in undernourished elderly people (Bastow *et al.*, 1983).

In recent years it has become evident that stressful conditions, which stimulate sympathoadrenal activity, play an important role in increasing morbidity and mortality from coronary heart and cerebrovascular disease. In heart failure, for example, reflexes fulfil a compensatory role so that maximum stimulation of the sympathetic system may already be occurring when exertion occurs, or even at rest. This is indicated by the high levels of plasma noradrenaline reflecting 'spill-over' from adrenergic nerve endings (Rector *et al.*, 1987). A permanent increase in sympathetic drive can, in the long run, be detrimental to the failing heart, not only because of vasoconstriction but because of receptor 'down regulation' (Bristow *et al.*, 1986). It has also been reported that autonomic neuropathy in adults predisposes to sudden unexplained death, the incidence of which is proportional to the severity of autonomic dysfunction (Page & Watkins, 1978). Most investigators have shown that though there may be sympathetic dysfunction, there is an increase in basal plasma catecholamine concentration with age. For many elderly, therefore, there may exist an altered sympathetic neuro-chemical and receptor response that may become detrimental in cold conditions (Collins, 1991).

## Conclusions

Most countries in cold and temperate regions show the same characteristic cold weather pattern of increased winter morbidity and mortality, but the seasonal swing is more pronounced in Great Britain than in many comparable countries in the same or higher latitudes. Epidemiological data indicate a seasonal effect with excess winter mortality being highest in the elderly group. The reasons seem obvious – cold weather and increased winter infections take their toll of the vulnerable elderly population – but the aetiology is not in fact fully established. There is a powerful relationship with ambient temperature; even a 1 °C fall in average winter temperature in Great Britain is sufficient to increase mortality rate significantly. An explanation is required, however, as to why many European countries with much colder winter climates and more pronounced cold temperature swings do not exhibit a greater excess mortality pattern. Increased use of central or district heating is one explanation, but the same excess winter mortality in the elderly is still reported to occur in those living in centrally heated dwellings in Great Britain.

Hypothermia contributes only a small part (perhaps 1%) of excess winter deaths according to recent statistics, and epidemics of influenza on average under 25%. There is an increase in mortality from respiratory disease in the winter months especially in elderly people suffering from chronic disabling respiratory illness. Circulatory diseases, principally ischaemic heart disease and cerebrovascular disease underlie the greater part of excess winter deaths.

In the British Isles, most of the variation can be accounted for by national mean winter temperatures, influenza epidemics, a secular trend showing a gradual reduction with time, and an effect of ageing in the demographic structure (Curwen, 1990/1). Various markers of socio-economic status (occupation, housing tenure, car access) strongly suggest a social class gradient with a lower excess winter mortality in Class I. Changes in respiratory mortality in winter appears to be most closely associated with influenza epidemics and improvements in home heating. Central heating in households in Great Britain has increased from 13% to 66% during the period 1964 to 1984. The immediate effect of outdoor cold may be a more important factor than indoor temperature in the aetiology of cardiovascular deaths in the elderly. On the whole, there appear to be characteristics of the British winter, lack of a winter culture, or housing conditions, that contribute to excessive morbidity and mortality. Cardiovascular deaths in old people in the winter may be due to a number of physiological factors including increased vulnerability to thrombosis, altered cardiovascular reflexes, cold-induced hypertension and age-related changes in autonomic nervous control.

### References

Allison, S.P. & Bastow, M.D. (1983). Undernutrition and femoral fracture. *Lancet*, i, 933–4.

Alderson, M.R. (1985). Season and mortality. *Health Trends*, **17**, 87–96.
Anderson, T.W. & Le Riche, W.H. (1970). Cold weather and myocardial infarction. *Lancet*, **i**, 291–6.
Bastow, M.D., Rawlins, J. & Allison, S.P. (1983). Undernutrition, hypothermia and injury in elderly women with fractured femur: an injury response to altered metabolism. *Lancet*, **i**, 143–5.
Boardman, B. (1993). Prospects for affordable warmth. In *Unhealthy Housing: Research, Remedies and Reform*, ed. R. Burridge & D. Ormandy, pp. 382–400. London: Spon.
Bristow, M.R., Ginsburg, R., Umans, V. *et al.* (1986). B1 and B2 adrenergic receptor sub-populations in non-failing and failing human ventricular myocardium: coupling of both receptor sub-types to muscle contraction and selective B1-receptor down-regulation in heart failure. *Circulation Research*, **59**, 297–309.
Bull, G.M. (1973). Meteorological correlates with myocardial and cerebral infarction and respiratory disease. *British Journal of Preventive and Social Medicine*, **27**, 108–13.
Bull, G.M. & Morton, J. (1975). Relationships of temperature with death rates from all causes and from certain respiratory and arteriosclerotic diseases in different age groups. *Age and Ageing*, **4**, 232–46.
Bull, G.M. & Morton, J. (1978). Environmental temperature and death rates. *Age and Ageing*, **7**, 210–24.
Burton, A.C. & Edholm, O.G. (1955). *Man in a Cold Environment*. London: Edward Arnold.
Collins, K.J. (1983). *Hypothermia the Facts*. Oxford: Oxford University Press.
Collins, K.J. (1986). Low indoor temperatures and morbidity in the elderly. *Age and Ageing*, **15**, 212–20.
Collins, K.J. (1987). Effects of cold on old people. *British Journal of Hospital Medicine*, **37**, 506–14.
Collins, K.J. (1991). The autonomic nervous system in old age. *Reviews in Clinical Gerontology*, **1**, 337–45.
Collins, K.J. & Exton-Smith, A.N. (1983). Thermal homeostasis in old age. *Journal of the American Geriatrics Society*, **31**, 519–24.
Collins, K.J., Easton, J.C., Belfield-Smith, H., Exton-Smith, A.N. & Pluck, R.A. (1985). Effects of age on body temperature and blood pressure in cold environments. *Clinical Science*, **69**, 465–70.
Collins, K.J., Sacco, P., Easton, J.C. & Abdel-Rahman, T.A. (1989). Cold pressor and trigeminal vascular reflexes in old age. In *Thermal Physiology 1989*, ed. J.B. Mercer, pp. 587–92. Amsterdam: Elsevier.
Curwen, M. (1990/1). Excess winter mortality: a British phenomenon? *Health Trends*, **22**, 169–75.
Curwen, M. & Devis, T. (1988). Winter mortality, temperature and influenza: has the relationship changed in recent years? *Population Trends*, **54**, 17–20.
Fox, A.J. & Goldblatt, P.O. (1982). *Longitudinal Study: Socio-demographic Mortality Differential*. London: HMSO (OPCS Series LS No. 1).
General Registrar Office (1840). *Second Annual Report of the Registrar General of Births, Deaths and Marriages in England for 1839*. London: HMSO.
General Registrar Office (1841). *Third Annual Report of the Registrar General of Births, Deaths and Marriages in England*. London: HMSO.

General Registrar Office (1874). *Deaths in London from the Cold Weather.* London: HMSO (Weekly Return No. 51, Vol. 35).

Guy, W.A. (1881). On temperature and its relation to mortality: an illustration of the application of numerical method to the discovery of truth. *Journal of the Statistical Society,* **44,** 235–62.

Kannel, W.B., Wolf, P.A., Castelli, W.P. & D'Agostino, R.B. (1987). Fibrinogen and risk of cardiovascular disease: the Framingham study. *Journal of the American Medical Association,* **258,** 1183–6.

Keatinge, W.R. (1986). Seasonal mortality among elderly people with unrestricted home heating. *British Medical Journal,* **293,** 732–3.

Keatinge, W.R., Coleshaw, S.R.K., Cotter, F. *et al.* (1984). Increases in platelet and red cell counts, blood viscosity, and arterial pressure during mild surface cooling: factors in mortality from coronary and cerebral thrombosis in winter. *British Medical Journal,* **289,** 1405–8.

Keatinge, W.R., Coleshaw, S.R.K. & Holmes, J. (1990). Changes in seasonal mortalities with improvement in home heating in England and Wales from 1964 to 1984. *International Journal of Biometeorology,* **33,** 71–6.

Lewis-Fanning, E. (1940). *A Comparative Study of the Seasonal Incidence of Mortality in England and Wales and in the United States of America.* London: HMSO (MRC Special Report Series No. 239).

McDowell, M. (1981). Long term trends in seasonal mortality. *Population Trends,* **26,** 16–19.

Nayha, S. (1984). The cold season and deaths in Finland. *Arctic Medical Research,* **37,** 20–4.

Office of Population Censuses and Surveys (1987). *Trends in Respiratory Mortality 1951–1975.* London: HMSO (OPCS Series DH 1 No. 7).

Page, M. & Watkins, P.J. (1978). Cardiorespiratory arrest and diabetic autonomic neuropathy. *Lancet,* i, 14–16.

Rector, T.S., Olivari, M.T., Levine, T.B. *et al.* (1987). Predicting survival for an individual with congestive heart failure using the plasma norepinephrine concentration. *American Heart Journal,* **114,** 148–52.

Rogot, E. & Padgett, S.J. (1976). Association of coronary and stroke mortality with temperature and snowfall in selected areas of the United States. *American Journal of Epidemiology,* **103,** 565–75.

Rose, G. (1966). Cold weather and ischaemic heart disease. *British Journal of Preventive and Social Medicine,* **20,** 97–100.

Royal College of Physicians of London (1966). *Report of Committee on Accidental Hypothermia.* London: Royal College of Physicians.

Sakamoto-Momiyama, M. (1977). *Seasonality in Human Mortality.* Tokyo: Tokyo University Press.

Stout, R.W. & Crawford, V. (1991). Seasonal variations in fibrinogen concentration among elderly people. *Lancet,* **338,** 9–13.

# 12 *Nutritional seasonality: the dimensions of the problem*

ANNA FERRO-LUZZI AND FRANCESCO BRANCA

The nutritional status of rural communities in developing countries is believed to undergo seasonal deterioration as a consequence of food shortages occurring during the months preceding the harvest (Teokul *et al.*, 1986). This is a time when food stores are at their lowest, often coinciding with the peak in demand for agricultural labour. This phenomenon is well known and the period is often referred to as the 'hungry season'. Seasonal epidemics of infectious and parasitic diseases may coincide with these periods and induce secondary forms of malnutrition. These affect mostly the vulnerable sectors of the community, namely children and pregnant women. Although primary and secondary seasonal malnutrition potentially overlap, this article will address solely the *primary* form. Moreover, only adult energy deficiency will be considered, leaving out of the discussion children, as well as any other form of nutritional deficiency.

In this article, the evidence for seasonal weight loss will be updated and a numerical estimate of the global dimensions of seasonal energy stress will be attempted. This will be performed by an integrated analysis of agro-climatic information provided by the seasonality index and of biological information derived from the anthropometric assessment of nutritional status.

## Seasonal body weight loss in the Third World

A review of the causes and occurrence of seasonality of energy metabolism in the Third World was presented at the first IDECG meeting in Guatemala about five years ago (Ferro-Luzzi *et al.*, 1987). On that occasion a tentative mapping of the occurrence and extent of seasonality in the world was obtained by combining absolute and relative seasonality indices in a single seasonality index (SI). This is based on rainfall, as well as on climate and soil characteristics. In low seasonality areas, rainfall is greater than 1000 mm/yr and is evenly distributed across the year; under these conditions, plants have optimal growth for at least 60% of the year. In

149

areas of moderate seasonality, rainfall ranges between 500 and 1000 mm/yr and clusters in two rainy seasons. The optimum plant growth occurs over 25–60% of the year. In high seasonality areas, rainfall is less than 500 mm/yr and is concentrated in a single rainy season; and only 25% of the year allows good plant growth. In high seasonality areas, agricultural labour is therefore forcibly concentrated in narrow slots of time, leading to bottlenecks in human energy demand. Also, food production is limited to specific times of the year, and this may periodically lead to absolute or relative reduction in the availability of energy and nutrients.

Although the SI is crude and qualitative, its application has revealed that, with the exclusion of desert areas, most regions of the world are characterized by a low or moderate degree of seasonality; only a few areas have the potential for serious seasonal energy stress (Ferro-Luzzi *et al.*, 1987). Thus seasonal bottlenecks of energy turnover might be less widespread than previously thought (Ferro-Luzzi *et al.*, 1987). Moreover, most human communities would have been exposed over generations to seasonally recurring cycles of negative energy balance; it is therefore likely that they would have by now developed strategies suitable to counter the threat periodically inflicted to their functional integrity and survival. Only under exceptional conditions, leading to failure of habitual avoidance strategies, would communities need to cope with energy stress and therefore to resort to adaptive strategies.

Avoidance strategies are mostly of an economic and socio-cultural nature; coping strategies, being of a behavioural, ergonomic and metabolic nature, have instead a biological basis. One of these biological manifestations consists of the mobilization of body energy stores, which may be taken as *the landmark* for the switching on of the coping strategy. Thus, weight loss represents the best and earliest indicator of the failure of the individual or community to avoid seasonal exposure to energy stress, and the amount of weight lost provides an excellent proxy measure of the severity of the exposure.

A previous review (Ferro-Luzzi *et al.*, 1987) pointed out that seasonal weight loss has been recorded in a number of communities in the Third World. The loss of weight rarely exceeds 5% in adults, women tending to have smaller losses than men. Five per cent corresponds to about 3 kg for a typical 60 kg man from the Third World.

Figs. 12.1 and 12.2 summarize studies showing weight loss in adults, incorporating new sets of data that have become available recently. Most studies are of a truly longitudinal nature, namely the same subjects were measured at two points of the year. The new data sets include studies conducted in different regions of Ethiopia (Ferro-Luzzi *et al.*, 1990), in India (McNeill *et al.*, 1988; Durnin *et al.*, 1990) Benin (Schultink *et al.*, 1990; DANA & GLTM, 1991), Zambia (Kumar, 1987) and Niger

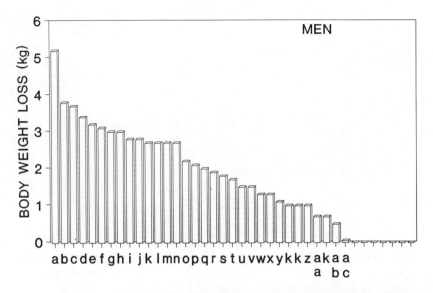

Fig. 12.1. Seasonal weight changes in adult men of several Third World populations. [a: Cameroon, 1980 (de Garine & Koppert, 1988); b: Burkina Faso (Ancey, 1974); c: Senegal (Gessain, 1978); d: Cameroon, 1976 (de Garine & Koppert, 1988); e: Mali, Fulbe (Hilderbrand, 1985); f: Niger (Loutan & Lamotte, 1984); g: Gambia (Fox, 1953); h: Gambia (Billewicz, 1982); i: Mali (Bénéfice & Chevassus-Agnès, 1985); j: Burkina Faso (Brun *et al.*, 1981); k: Benin, several communities (DANA & GLTM, 1991); l: Mali, Rimaibe (Hilderbrand, 1985); m: Botswana (Wilmsen, 1978); n: Senegal, Ferlo (Bénéfice & Chevassus-Agnès, 1985); o: Senegal (Rosetta, 1986); p: Kenya (Little, 1989); q: Burma (Tin-May-Than & Ba-Aye, 1985); r: Zambia (Kumar, 1987); s: Peru (Leonard & Thomas, 1989); t: Niger (Reardon, 1989); u: Ethiopia, Sidamo (Branca *et al.*, 1989); v: Zaire, Ntombe (Pagezy, 1984); w: Ethiopia, Shoa (Ferro-Luzzi *et al.*, unpublished); x: Papua New Guinea (Spencer & Heywood, 1983); y: Zaire, Walese (Bailey & Peacock, 1988); z: Bangladesh, west central (Abdullah & Wheeler, 1985); aa: India, Karnataka (Norgan *et al.*, 1986); ab: India, Tamil Nadu (McNeill *et al.*, 1988); ac: Bangladesh, north (Abdullah & Wheeler, 1985)].

(Reardon, 1989) in addition to several studies carried out in pastoralist and hunter–gatherer groups in Cameroon (de Garine & Koppert, 1988), Botswana (Wilmsen, 1978), Mali (Hilderbrand, 1985) and Kenya (Little, 1989). There is also one study on pygmies (Bailey & Peacock, 1988) and one on Peruvian highlanders (Leonard & Brooke Thomas, 1989). Sample sizes vary, ranging from only a few subjects, as in the case of the Burmese farmers studied by Tin-May-Than & Ba-Aye (1985), to in excess of two thousand, as in the study of Kenyan women by Cogill (Cogill, 1987). As can be seen in Figs. 12.1 and 12.2, the previously defined range of weight loss is basically confirmed for all but two groups, one in Cameroon (de Garine & Koppert, 1988), the other in the Gambia (Lawrence *et al.*, 1989). However, when studied in different years, neither of these two groups experienced the same high level of seasonal nutritional stress.

152     *Anna Ferro-Luzzi and Francesco Branca*

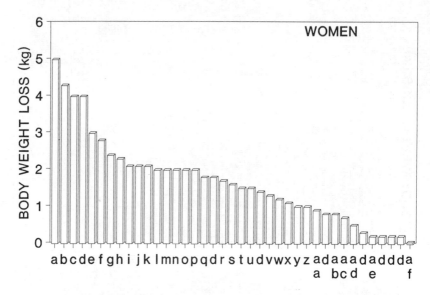

Fig. 12.2. Seasonal weight changes in adult women of several Third World populations. [a: Gambia (Lawrence *et al.*, 1989); b: Cameroon, 1980 (de Garine & Koppert, 1988); c: Kenya (Little, 1989); d: Benin (DANA & GLTM, 1991); e: Papua New Guinea (Crittenden & Baines, 1986); f: Zambia (Kumar, 1987); g: Niger (Loutan & Lamotte, 1984); h: Mali, Fulba (Hilderbrand, 1985); i: Mali (Bénéfice & Chevassus-Agnès, 1985); j: Senegal, Ferlo (Bénéfice & Chevassus-Agnès, 1985); k: Cameroon, 1976 (de Garine & Koppert, 1988); l: Gambia (Billewicz, 1982); m: Gambia (Prentice *et al.*, 1981); n: Burkina Faso (Ancey, 1974); o: Bangladesh (Chen *et al.*, 1979); p: Mali, Rimaibe (Hilderbrand, 1985); q: Ethiopia, Shoa (Ferro-Luzzi *et al.*, unpublished); r: Papua New Guinea (Spencer & Heywood, 1983); s: Ethiopia, Sidamo (Ferro-Luzzi *et al.*, 1990); t: Senegal (Rosetta, 1986); u: Bangladesh, north (Abdullah & Wheeler, 1985); v: Ethiopia, Sidamo (Branca *et al.*, 1989); w: Peru (Leonard & Brooke Thomas, 1989); x: Benin, Mono (Schultink *et al.*, 1990); y: Bangladesh, west central (Abdullan & Wheeler, 1985); z: Zaire, Walese (Bailey & Peacock, 1988); aa: Kenya, South Nyanza (Cogill, 1987); ab: Zaire, Ntombe (Pagezy, 1984); ac: Burkina Faso (Bleiberg *et al.*, 1980); ad: India, South (Durnin *et al.*, 1990); ae: India, Karnataka (Norgan *et al.*, 1989); af: India, Tamil Nadu (McNeill *et al.*, 1988).]

In a large variety of situations, exposure to seasonal energy stress translates into a periodic loss of body weight ranging from a minimum of 1 kg to a maximum of 5 kg. The highest losses are recorded in some communities as a result of the superimposition of long-term climatic cycles on yearly rainfall fluctuation.

## Biological and functional significance of weight loss

The withdrawal of energy from body fat stores is a physiological mechanism that allows immediate needs to be met. For an individual losing 4 kg over one to two months, a total of about 117 MJ (28 000 kcal) would be

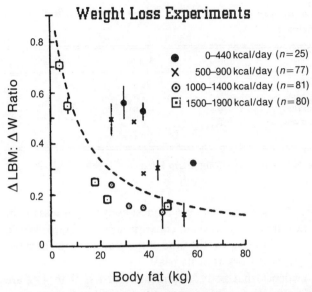

Fig. 12.3. Proportion of body weight lost as lean tissue at different levels of body fat stores. The curve has been drawn from the results of weight loss experiments where caloric intake was only moderately reduced, and was not inferior to 1000 kcal. ΔLBM : ΔW = change in body mass: change in body weight. (Modified from Forbes, 1989.)

made available by this mechanism, at the rate of about 1.9 to 3.8 MJ (450 to 900 kcal) daily, to cover the discrepancy between energy expenditure and intake. This represents a turnover of less than 1% of total body energy stores for an average adult male. Energy comes from fat mobilization, and from a modest loss of lean tissue. Data from Forbes (1989), summarized in Fig. 12.3 show that, unless daily energy intake is well below 4.2 MJ (1000 kcal) a day, the proportion of lean tissue lost is inversely proportional to the initial size of energy stores. In a well-fed person losing 4 kg of body weight, this would entail a loss of about 400 g of the supporting matrix of the adipose tissue, while another 400 g would be muscle tissue. While 400 g of muscle tissue represents a negligible proportion in a well-nourished person, for a man in the Third World community, whose body energy stores are much lower to start with, the picture may be quite different. The body weight of an African or an Indian peasant is about 60 kg and, at 10% body fat, his total energy reserve would be 6 kg of lipids. Under such marginal conditions, the proportional loss of lean tissue would be much larger and might account for more than 60% of the weight lost. Accordingly, the hypothetical Third World farmer losing 4 kg body weight, would lose about 1.5 kg fat but also 2.5 kg of lean tissue. As the latter would be reintegrated with difficulty, given the quality of the diet prevalent in rural areas of developing countries, this process could lead to progressive

Table 12.1. *Metabolic responses to seasonal weight change of three groups of Third World women*

|  | India[a] | Benin[b] | Ethiopia[c] |
|---|---|---|---|
| BMI at harvest (kg/m$^2$) | 18.0 | 20.8 | 18.4 |
| Body weight change from harvest to lean season (kg) | $-0.3$ | $-1.0$ | $-1.6$ |
| BMR change from harvest to lean season (%) | $-5$ | 0 | $-7$ |

[a]Durnin *et al.* (1990); [b]Schultink *et al.* (1990); [c]Ferro-Luzzi *et al.* (1990).

wasting. Therefore, it is possible that even modest body weight loss might have important health and functional consequences in Third World communities; more so if it occurs at the time of the year when the demand for high intensity physical work is at its peak.

There is good evidence that body weight losses of less than 4 kg are capable of inducing a measurable metabolic response, if the loss is sustained by individuals with impaired nutritional status (Ferro-Luzzi, 1990). Beninese and Ethiopian women have been found to lose 1 to 1.5 kg body weight during the lean season, but while there was no change in the basal metabolic rate (BMR) of the plump Beninese women with a body mass index (BMI) of 20.8 kg/m$^2$, the BMR of the thin Ethiopian women with a BMI of 18.4 kg/m$^2$ dropped significantly (Table 12.1; Ferro-Luzzi, 1990). In another study, Indian women with a BMI of 18 kg/m$^2$ experienced a reduction in BMR during the lean season despite the fact that their weight loss was negligible (0.3 kg). Other authors (McNeill *et al.*, 1988) did not find evidence of a similar metabolic response in Indian women losing 0.5 kg in weight, but these women had a mean BMI of 19.1 kg/m$^2$. The only other study where a decrease in resting metabolic rate was observed was one on Gambian women with a BMI of 20.5 kg/m$^2$ and a weight loss of 5 kg (Lawrence *et al.*, 1989). These results, while confirming that weight loss is a valid marker of exposure to energy stress, suggest that the biological significance of similar losses may vary as a function of the status of the body energy stores.

It is therefore interesting to note that the amount of seasonal weight loss appears to be inversely related to fatness (as measured by BMI), prior to seasonal stress. Ethiopian women with BMIs in the two lowest quintiles of distribution, $<17.5$ kg/m$^2$ and $17.6–18.5$ kg/m$^2$ respectively, lost 0.5% to 2.5% of their body weight, while those in the higher quintiles had a weight loss that was greater than 3% (Ferro-Luzzi *et al.*, 1990). A similar finding was noted in Benin, where women with a BMI above 23 kg/m$^2$ lost

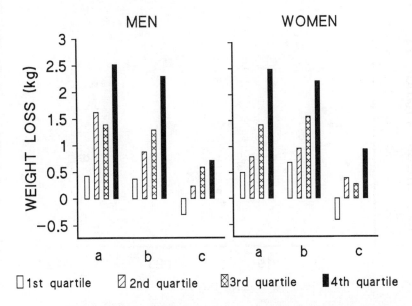

Fig. 12.4. Seasonal body weight loss of men and women in the Third World, by quartile of BMI. The data sets derive from three longitudinal studies conducted in (a) Sidamo, Ethiopia (Branca *et al.*, 1989), (b) Southern Shoa, Ethiopia (Ferro-Luzzi *et al.*, unpublished), and (c) Karnataka, India (Norgan *et al.*, 1989).

significantly more weight than those with a BMI less than $18 \, \text{kg/m}^2$ (Schultink *et al.*, 1990). The same phenomenon had already been reported in Zaire (Pagezy, 1984) and was later confirmed by three further studies conducted in the Sidamo and Southern Shoa Region in Ethiopia (Branca *et al.*, 1989) and in Karnataka in India (Norgan *et al.*, 1989). Fig. 12.4 shows consistently larger seasonal weight loss of individuals with higher initial BMI.

The biological basis of this phenomenon is uncertain, and there are many confounding socio-economic factors to the notion that the greater weight loss of the more plump members of a community has a purely biological basis. For example, the poorest peasants may unwillingly spare their energy during the lean pre-harvest time of the year, given that their smaller plots of land require less physical labour. They would thus lose less body weight than the richer peasants, whose larger farms demand more human physical labour. Although this may prove to be the case, it is interesting to note that the relationship of BMI-to-weight-loss also holds within socio-economic classes. Fig. 12.5 shows the seasonal changes in body weight of the richest and the poorest peasants in a rural community in Sidamo, plotted against BMI. As expected, the BMI of the poorer peasants is lower than that of the

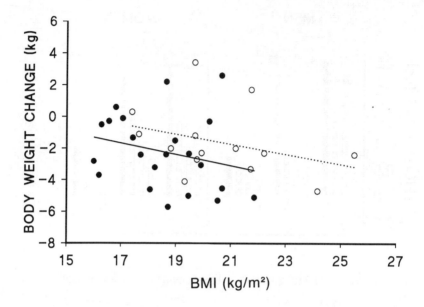

Fig. 12.5. Seasonal changes in body weight and BMI observed in a longitudinal study conducted in Ethiopia (Sidamo province, unpublished data). Subjects have been distributed on the basis of their socio-economic status. The two extreme deciles of the distribution are here presented as poor (solid circle) and rich (blank circle) peasants. Each circle corresponds to one individual. Linear regression lines have been calculated separately for the poor (solid line) and the rich (broken line) individuals.

richer ones. The two slopes are similar and reveal a similar tendency of larger weight changes at higher BMI values. The only difference is that, for the same BMI, poor peasants tend to lose less body weight than rich ones. However, given the small number of subjects falling in these two extreme socio-economic groups, the differences are not statistically significant.

Also, the inter-country comparison of weight loss and BMI points towards such a relationship. The left-hand panel of Fig. 12.6 shows the mean weight loss reported in the literature for men from several communities. Among farmers, the communities with lower mean BMI tend to lose less body weight. All six communities of pastoralists suffer more sustained losses, 2 to 5 kg, and their BMI/weight loss relationship lies outside that of the farmers. They have been excluded and the regression line is drawn using data from farming communities only. Cameroon peasant farmers, who were studied in a year of severe drought, are also excluded. Among the women (right-hand panel of Fig. 12.6), the relationship is less straightforward and appears to be better described by a cubic equation, or by the

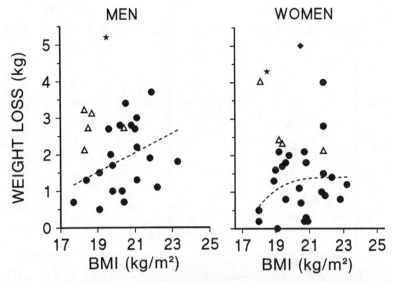

Fig. 12.6. Mean seasonal body weight changes by BMI among Third World peasants: a cross-country analysis of the communities illustrated in Figs 12.1 and 12.2. The broken line denotes the best fit for the farmers (solid circles). Pastoralists (triangles), Cameron farmers during a drought (stars) and pregnant Gambian women (diamond) have not been included in the regression.

intersection of two regression lines, a steeper one for BMI below 19 kg/m², a second flatter one for BMI above 19 kg/m². Groups studied under particularly stressful environmental conditions (Cameroon and the Gambia) and pastoralists are shown but not included.

The consistency of such observations under real life conditions within and across communities strongly suggests the existence of a biological constraint to the amount of fat that may be removed from the body as a proportion of total body fat. It may be argued that when weight losses exceed a given threshold of this proportion, functional and health deterioration would occur thus defeating the coping mechanisms.

There is little documentation in the literature on the functional correlates of seasonal weight losses, and none of it takes into account the BMI–weight loss relationship. Thus in any estimate of the world-wide extent of seasonal energy stress exposure, a degree of speculation is inevitable. In Fig. 12.7, which displays the same groups as in Fig. 12.6, a new regression line has been drawn, deliberately retaining only communities experiencing the largest weight loss at each BMI level. The two communities studied in drought periods were excluded because of the uniqueness of their situation. The pastoralists and hunter–gatherers have also been excluded from the analysis because of the many peculiarities of their environment, social

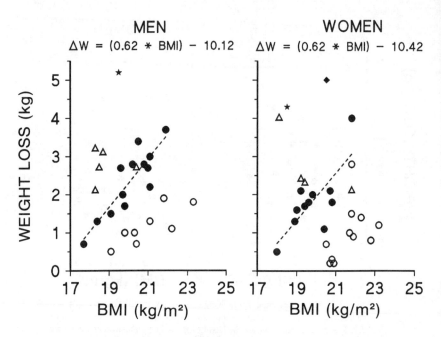

Fig. 12.7. Seasonal body weight change by BMI among Third World peasants (see Figs 12.1 and 12.2). The regression lines have been drawn on the communities exhibiting the maximum weight loss at each BMI level (solid circles). The pastoralists (triangles), the drought-stricken Cameron farmers (stars) and the pregnant Gambian women (diamond) have not been included.

systems and dietary patterns that set them apart from agriculturalists. The basis of this selection is the assumption that the regression line drawn here identifies the maximum weight loss that can be sustained by a community at any given mean BMI without unduly compromising the functional and metabolic integrity of the individuals within it. This line might be called the 'maximal stress line'. Communities plotted to the left of the line are considered at risk of a costly deterioration of their nutritional status and functional integrity, while those falling to the right might be considered reasonably safe. In other words, it may be argued that this line defines the biological resistance of the organism to removing fat from the body, resistance that increases in inverse relation to the amount of fat present in the body. The line has a plausible zero loss intercept at BMI 16.5 kg/m² for males and 16.8 kg/m² for females. These values are close to the proposed cut-off point for third grade chronic energy deficiency (Ferro-Luzzi *et al.*, 1992). The regression line for women has a higher intercept than for men, possibly reflecting protection against the extra energy stress associated with reproduction.

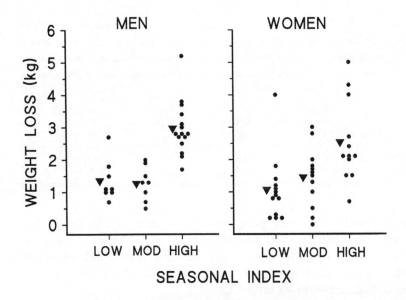

Fig. 12.8. Mean seasonal body weight changes observed in areas characterized by low, moderate and high Seasonality Index. The dots represent the mean for each study community (see Figs 12.1 and 12.2). The triangles indicate the average seasonal body weight loss at each Seasonality Index (see text for explanation of the Seasonality Index).

### The world dimensions of nutritional seasonality

The three sets of information needed to attempt an estimate of the dimension of world-wide nutritional seasonality may now be established, although it is acknowledged that the central assumptions are somewhat speculative and further research is required to put them on a firmer footing. Seasonality indices of the various regions of the world constitute the first set of information. The second set, the intensity of energy stress associated with each value of the seasonality index, is derived from data in the literature. The third set, the minimum BMI at which the expected energy stress can be met without compromising the health and the function of the community, is calculated using the 'maximal stress line' equation.

Fig. 12.8 shows the distribution of weight loss recorded in regions classified as being of low, moderate and high seasonality. Data are presented separately for men and women. The losses in regions with low and moderate seasonality are similar and have been combined; in these regions, both men and women suffer an average weight loss of 1.5 kg. In areas with high seasonality men lose 3 kg and women 2.5 kg.

The BMI value at which these weight losses may occur without seriously

Table 12.2. *Body weight losses expected to occur in low/moderate and high seasonality areas and cut-off points of BMI above which the specified weight losses are assumed to take place without incurring functional impairment*

|  | Men | | Women | |
|---|---|---|---|---|
| Seasonality index | Expected weight loss (kg) | Minimum BMI (kg/m$^2$) | Expected weight loss (kg) | Minimum BMI (kg/m$^2$) |
| Low/moderate | 1.5 | 18.7 | 1.5 | 19.2 |
| High | 3.0 | 21.2 | 2.5 | 20.8 |

compromising health and performance may now be calculated. The maximal stress line equation gives BMI values of 18.7 kg/m$^2$ for men and 19.2 kg/m$^2$ for women for communities exposed to low/mild seasonality energy stress, and a BMI of about 21 kg/m$^2$ for both men and women in areas of high seasonality (Table 12.2).

With these data it is now possible to proceed to evaluate the number of people in the world who fall on the wrong side of the maximal stress line. This analysis is limited by the scarcity of data available to calculate the SI; it is also limited in the number of representative data sets of adult weights and heights for most developing countries.

Fig. 12.9 shows the distribution of low, moderate and high Seasonality Index in 19 African countries for which both agro-climatic and anthropometric data were available. As mentioned earlier, desert areas are not included. The total population of these 19 countries, as given by the UN Department of International Economic and Social Affairs (UN, 1980), is 136 million. Literature sources have provided mean BMI values (Eveleth & Tanner, 1976, 1990; Ferro-Luzzi *et al.*, 1992; James, pers. commun.). The regions where the mean BMI is below the value compatible with the weight loss expected at the given SI are shown as superimposed darker areas. So, for example, the adult males in the western Sahelian regions are presumed to drop their body weights by 3 kg at every pre-harvest season. Given their mean BMI of 20, this weight loss is expected to induce metabolic and functional impairments. On this basis, it appears that only Burkina Faso, the Gambia, Sudan, Ethiopia and Kenya belong to the high risk category.

The actual number of people thus exposed has been calculated for these countries, taking two assumptions into account. Firstly, that only the rural sectors of the population are exposed to seasonal energy stress; and secondly, that the mean BMI figure obtained from literature sources had, unless otherwise specified, a 10% coefficient of variation. Thus, it is estimated that 46% of rural men and 36% of rural women living in African

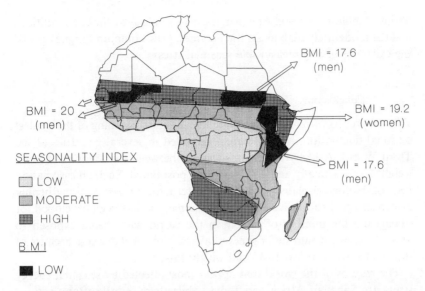

Fig. 12.9. Map of the Seasonality Index in some countries of Africa. Areas where the mean BMI of the population is below the threshold deemed compatible with the body weight loss expected on the basis of the Seasonality Index are shaded in black.

countries with high SI (involving a weight loss of 2.5 to 3.0 kg) are at risk of functional and metabolic impairment because of their low BMI. In areas with low/moderate seasonality, 20% of rural men and 20% of rural women are estimated to run a similar risk. In these 19 African countries, the total estimate is of 30 million rural adults suffering each year from exposure to energy stress to a degree that exceeds their physiological coping capacity.

A very rough attempt to make a similar calculation for other regions of the world has also been carried out. Seasonal nutritional exposure is likely to be experienced in low-BMI rural populations in Asia. In India, mean BMIs of 18.8 for men and 18.4 for women (Norgan *et al.*, unpublished data), make these populations particularly vulnerable even to modest shifts of their body weight. Thus, even assuming a low/moderate SI for the whole country, calculations reveal that the vast majority of rural women (82%) and almost half of rural men (45%) are at high risk of energy nutritional stress because of their low BMI. Given the large population of the Indian subcontinent, these percentages translate into about 200 million individuals. The situation in China is rather different. Although seasonality indices could not be calculated for this region, the relatively high BMI ($21.1\,\mathrm{kg/m^2}$ for men and $21.6\,\mathrm{kg/m^2}$ for women) and its rather tight distribution (James, pers. commun.) makes it unlikely that more than a few per cent of the population is at risk. However, as China has the largest population in the world, even this small proportion translates to a figure of about 80 million. In Latin America, the favourable agroclimatic conditions and the moderately high mean BMI (James, pers. commun.) argue against the existence of any measurable seasonal stress.

## Conclusions

This chapter has attempted to assess the biological meaning of the modest seasonal fluctuation in body weight recorded in several countries of the Third World. A likely inverse association between the amount of body weight lost seasonally and BMI has been postulated. Such an association rests on biological plausibility, but its interpretation needs strengthening evidence. From the above assumptions, a method has been developed for calculating the number of people in the world who, being exposed to seasonal stress, would suffer functional and biological damage because of the combination of low BMI and weight loss.

The regions of the world that appear most affected by seasonal energy stress are Sahelian Africa and India, while there is little if any serious concern for other developing regions and countries. This very rough calculation has not been extended to all countries of the Third World for lack of agroclimatic and/or anthropometric data. For the countries for which the calculation was possible, a global figure of about 300 million is reached. These are individuals who should be considered at risk of functional and metabolic impairment following their exposure to seasonal bottlenecks in energy turnover that exceed the body's physiological tolerance. The cost in terms of human capital and productive agricultural potential of seasonal energy stress may therefore be higher than suggested by the modesty of the recorded body weight losses.

## Acknowledgement

Part of this work was supported by the financial contribution of the Directorate General XII for Science Research and Development of the Commission of the European Communities (Research grant TS2-0154-I). The authors wish also to thank Ms E. Toti for skillful and careful help in the preparation of the manuscript, and Ms P. Ferranti and Ms G. Alicino for the preparation of the figures and tables.

## References

Abdullah, M. & Wheeler, E.F. (1985). Seasonal variations, and the intra-household distribution of food in a Bangladeshi village. *American Journal of Clinical Nutrition*, **41**, 1305–13.

Ancey, G. (1974). *Facteurs et systèmes de production dans la société Mossi d'aujourd'hui. Migration-travail-terre et capital.* Ouagadougou: ORSTOM.

Bailey, R.C. & Peacock, N.R. (1988). Efe Pygmies of northeast Zaire: subsistence strategies in the Ituri forest. In *Coping with Uncertainty in Food Supply*, ed. I. de Garine & G.A. Harrison, pp. 88–117. Oxford: Oxford University Press.

Bénéfice, E. & Chevassus-Agnès, S. (1985). Variations anthropométriques saisonnières des adultes appartenant à deux populations différentes de l'Afrique de l'Ouest. *Revue d'Epidémiologie et Santé Publique*, **33**, 150–60.

Billewicz, W.Z. (1982). A birth-to-maturity longitudinal study of heights and weights in two West African (Gambian) villages, 1951–1975. *Annals of Human Biology*, **9**, 309–20.

Bleiberg, F.M., Brun, T.A. & Goihman, S. (1980). Duration of activities and energy expenditure of female farmers in dry and rainy seasons in Upper-Volta. *British Journal of Nutrition*, **43**, 71–82.

Branca, F., Pastore, G., Demissie, T. & Ferro-Luzzi, A. (1989). Biological impact of seasonality in Ethiopia. Paper presented at the IFPRI workshop 'Seasonality in agriculture. Its nutritional and productivity implications'. Washington, January 26–7.

Brun, T., Bleiberg, F. & Goihman, S. (1981). Energy expenditure of male farmers in dry and rainy seasons in Upper-Volta. *British Journal of Nutrition*, **45**, 67–75.

Chen, L.C., Alauddin Chowdhury, A.K.M. & Huffman, S.L. (1979). Seasonal dimensions of energy protein malnutrition in rural Bangladesh: the role of agriculture, dietary practices, and infection. *Ecology of Food and Nutrition*, **8**, 175–87.

Cogill, B. (1987). *Seasonal Influences in South-west Kenya.* Washington: IFPRI Research Report.

Crittenden, R. & Baines, J. (1986). The seasonal factors influencing child malnutrition on the Nembi plateau, Papua New Guinea. *Human Ecology*, **14**, 191–223.

Direction de l'Alimentation et de la Nutrition Appliquée & Groupe Laics Tiers Mond (1991). *Le système alimentaire et l'état nutritionnel dans 11 sous-préfectures rurales du Bénin.* Rome: Porto Novo.

de Garine, I. & Koppert, G. (1988). Coping with seasonal fluctuations in food supply among savanna populations: the Massa and Mussey of Chad and Cameroon. In *Coping with Uncertainty in Food Supply*, ed. I. de Garine & G.A. Harrison, pp. 210–59. Oxford: Oxford University Press.

Dugdale, A.E. & Payne, P.R. (1987). A model of seasonal changes in energy balance. *Ecology of Food and Nutrition*, **19**, 231–45.

Durnin, J.V.G.A., Drummond, S. & Satyanarayana, K. (1990). A collaborative EEC study on seasonality and marginal nutrition: the Glasgow–Hyderabad (S. India) study. *European Journal of Clinical Nutrition*, **44** (Suppl. 1), 19–29.

Eveleth, P.B. & Tanner, J.M. (1976). *Worldwide Variation in Human Growth.* Cambridge: International Biological Programme & Cambridge University Press.

Eveleth, P.B. & Tanner, J.M. (1990). *Worldwide Variation in Human Growth.* Cambridge: Cambridge University Press.

Ferro-Luzzi, A. (1990). Seasonal energy stress in marginally nourished rural women: interpretation and integrated conclusions of a multicentre study in three developing countries. *European Journal of Clinical Nutrition*, **44** (Suppl. 1), 41–6.

164     *Anna Ferro-Luzzi and Francesco Branca*

Ferro-Luzzi, A., Pastore, G. & Sette, A. (1987). Seasonality in energy metabolism. In *Chronic Energy Deficiency: Consequences and Related Issues*, ed. B. Schürch & N.S. Scrimshaw, pp. 37–58. Lausanne: IDECG.

Ferro-Luzzi, A., Scaccini, C., Taffese, S., Aberra, B. & Demeke, T. (1990). Seasonal energy deficiency in Ethiopian rural women. *European Journal of Clinical Nutrition*, **44** (Suppl. 1), 7–18.

Ferro-Luzzi, A., Sette, S., Franklin M. & James, W.P.T. (1992). A simplified approach to assessing adult chronic energy deficiency. *European Journal of Clinical Nutrition*, **46**, 173–86.

Forbes, G.B. (1989). Changes in body composition. In *Report of the 98th Ross Conference on Pediatric Research*, pp. 112–18. Columbus: Ross Laboratories.

Fox, R.H. (1953). A study of the energy expenditure of Africans engaged in various activities, with special references to some environmental and physiological factors which may influence the efficiency of their work. PhD Thesis. University of London (cit. in Dugdale & Payne, 1987).

Gessain, M. (1978). Poids individuels saisonniers chez les Bassari du Sénégal Oriental. *Bulletin et Mémoires de la Société d'Anthropologie de Paris*, **5**, 149–55.

Hilderbrand, K. (1985). Assessing the components of seasonal stress amongst Fulani of the Seno-Mango, Central Mali. In *Population, Health and Nutrition in the Sahel. Issues in the Welfare of Selected West African Communities*, ed. A.G Hill, pp. 208–87. London: Routledge & Kegan Paul.

Kumar, S. (1987). Effect of seasonal food shortage on agricultural production in Zambia. Paper presented at the joint IFPRI–FSG workshop on 'Issues in food security'. Oxford, July 7–9, 1987.

Lawrence, M., Lawrence, F., Cole, T.J., Coward, W.A., Singh, J. & Whitehead, R.G. (1989). Seasonal pattern of activity and its nutritional consequence in the Gambia. In *Causes and Implications of Seasonal Variability in Third World Agriculture: The Consequences for Food Security*, ed. D. Sahn. Baltimore: Johns Hopkins University Press.

Leonard, W.R. & Thomas B.R. (1989). Biosocial responses to seasonal food stress in highland Peru. *Human Biology*, **61**(1), 65–85.

Little, M.A. (1989). Human biology of African pastoralists. *Yearbook of Physical Anthropology*, **32**, 215–47.

Loutan, L. & Lamotte, J.M. (1984). Seasonal variations in nutrition among a group of nomadic pastoralists in Niger. *Lancet*, **i**, 945–7.

McNeill, G., Payne, P.R., Rivers, J.P.W., Enos, A.M.T., De Britto, J. & Mukarji, D.S. (1988). Socio-economic and seasonal patterns of adult energy nutrition in a south Indian village. *Ecology of Food Nutrition*, **22**, 85–95.

Norgan, N.G., Shetty, P., Baskaran, T. & Ferro-Luzzi, A. (1989). Seasonality in a Karnataka (South India) agricultural cycle: background and body weight changes. Paper presented at the KFPRI workshop 'Seasonality in agriculture. Its nutritional and productivity implications'. Washington, January 26–7.

Pagezy, H. (1984). Seasonal hunger as experienced by the Oto and the Twa women of a Ntomba village in the Equatorial forest (Lake Tumba, Zaire). *Ecology of Food and Nutrition*, **15**, 13–27.

Prentice, A.M., Whitehead, R.G., Roberts, S.B. & Paul, A.A. (1981). Long-term energy balance in child-bearing Gambian women. *American Journal of Clinical Nutrition*, **34**, 2790–9.

Reardon, T. (1989). Seasonality in household transactions in western Niger. Paper presented at the IFPRI workshop 'Seasonality in agriculture. Its nutritional and productivity implications'. Washington, January 26–7.

Rosetta, L. (1986). Sex differences in seasonal variations of the nutritional status of Serere adults in Senegal. *Ecology of Food and Nutrition*, **18**, 231–44.

Schultink, W.J., Klaver, W., Van Wijk, H., Van Raaij, J.M.A. & Hautvast, J.G.A.J. (1990). Body weight changes and basal metabolic rates of rural Beninese women during seasons with different energy intakes. *European Journal of Clinical Nutrition*, **44** (Suppl. 1), 31–40.

Spencer, T. & Heywood, P. (1983). Seasonality, subsistence agriculture and nutrition in a lowland community of Papua New Guinea. *Ecology of Food and Nutrition*, **13**(4), 221–9.

Teokul, W., Payne, P.E. & Dugdale, A. (1986). Seasonal variations in nutritional status in rural areas of developing countries. Review of the literature. *Food and Nutrition Bulletin*, **8**, 7–10.

Tin-May-Than & Ba-Aye (1985). Energy intake and energy output of Burmese farmers at different seasons. *Human Nutrition: Clinical Nutrition*, **39C**, 7–15.

UN Department of International Economic and Social Affairs (1980). *Patterns of Urban and Rural Population Growth*. New York.

Wilmsen, E.N. (1978). Seasonal effects of dietary intake on Kalahari San. *Federation Proceedings*, **37**(1), 65–72.

# 13 Seasonal variation in nutritional status of adults and children in rural Senegal

KIRSTEN B. SIMONDON, ERIC BÉNÉFICE,
FRANÇOIS SIMONDON, VALÉRIE DELAUNAY AND
ANOUCH CHAHNAZARIAN

Seasonal variation in nutritional status of adults and children in rural West Africa has been described previously, both for farmers (Gessain, 1978; Prentice et al., 1981; Bénéfice & Chevassus-Agnès, 1985; Rosetta, 1986; Schultink et al., 1990) and for pastoralists (Bénéfice et al., 1984; Loutan & Lamotte, 1984).

The critical period during which nutritional status is impaired generally occurs around the end of the rainy season for farmers and around the end of the dry season for pastoralists. In both cases, the determinant invoked is the concomitance of increased physical activity in adults and seasonal food shortage.

This chapter describes seasonal variation in the nutritional status of adults and children in two agricultural societies in Senegal, West Africa. Socioeconomic characteristics of the two populations are very similar: both have low educational levels, both derive their subsistence from small scale farming while monetary income is generated from the sale of specific crops and through remittances of emigrees. Both areas are characterized by one marked rainy season per year. However, the major difference between the two study areas resides in the total independance from rainfall recently achieved in one of the areas (the Senegal River Valley) through massive irrigation works.

## Data and methods

### The Senegal River Valley study area

The Senegal River flows in the north of the country, forming the border with Mauritania (Fig. 13.1). Major developments have taken place since 1981 under the direction of *Office de Mise en Valeur du Fleuve Sénégal*

166

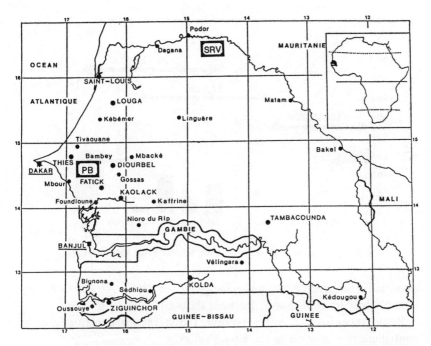

Fig. 13.1. Map of Senegal showing the two study areas, the Senegal River Valley (SRV) and the Peanut Basin (PB).

(OMVS). A reservoir-barrage has been built upstream at Manantali in Mali and an 'antisalt-barrage' constructed in the Senegal Estuary near St. Louis on the Atlantic Coast.

The aim is to replace the two main traditional farming methods, the *Diery* cultivation of millet during the rainy season and the *Walo* cultivation of sorghum in small basins flooded at the end of the rainy season (flood-recession culture), with irrigated agriculture across the year. Annual rainfall has varied between 97 and 320 mm during the last 10 years.

The nutritional study in the Senegal River Valley (SRV) was undertaken in January 1990 as a part of a multidisciplinary study of the impact of the agricultural development project on the health status of the population. The study area consists of three villages surrounding an irrigated area of 580 ha.

Irrigation started in June 1989 and the first rice harvest took place in December 1989. The population of 3300 belong to the Toucouleur ethnic group; they are sedentary farmers and fervent Muslims. Toucouleurs belong to the same ethnic group as Peuls, the *Hal puular* group. They have the same language, but while Peuls are nomadic pastoralists, Toucouleurs seem to have settled around the time of the twelfth century and are now

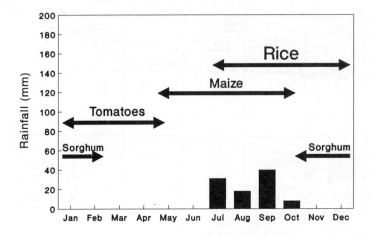

Fig. 13.2. Agricultural calendar in the Senegal River Valley in 1991.

essentially farmers. Peuls are widespread in West Africa with large communities in Mali, Niger, Burkina Faso and Cameroun.

Remittances from emigrated relatives (younger men) constitute a major contribution to income in the River Valley.

Fig. 13.2 presents the 1991 agricultural calendar. The major crop for most families was rice, which was sowed directly in the irrigated fields (without bedding out) in July, cleared during the following months and harvested in December. During the last month before harvest, children keep birds away from the rice fields. Maize was cultivated between May and October, mainly as a food crop, while tomatoes were grown, by the women only, between January and May as a cash crop (for canning). Sorghum was grown in traditional fields between October and February by a few families who had no access to the irrigated fields.

### Subjects

A random sample consisting of 110 compounds was selected using a cluster design, whereby the probability for a given compound being selected was proportional to the number of members in the compound. The aim was to select one-third of the population in the nutritional study.

From that sample, all children aged less than 5 years ($n = 330$) together with their mothers ($n = 197$) and all men aged 20–60 years ($n = 101$) were included.

Pregnant women were excluded from the analysis of anthropometric data, while lactating women were included independently of lactation stage.

## Methods

### Anthropometry

Anthropometric measurements were taken on all subjects present in February, June, October and December 1991 and in February 1992. Mean numbers of subjects were 66 infants, 300 preschool children, 164 women and 88 men. A smaller subsample of children (n = 145) and mothers (n = 76) were measured in April 1991, a few days after the end of Ramadan, the Muslim month of fasting.

Recumbent length was measured on children below the age of 24 months using a locally made wooden board (precision: ±0.1 cm). The same board was used for the measurements of height in older children. Children below approximately 10 kg were weighed naked on a SECA scale (precision: ±10 g) while heavier children and adults were weighed on an electronic Téfal scale (precision: ±200 g). They wore light clothing and no adjustment for the weight of clothes was made. Height of adults was measured using a Harpenden anthropometer (precision: ±0.1 cm), removing headclothes prior to measurement. Scales were calibrated every morning using standard weights.

Measurements were taken by a mobile team in each compound. In periods of intensive farming work, men were weighed in the fields. Birth dates of children were determined according to the Muslim calendar through interviews of their mothers. The Muslim birth dates were then converted into Gregorian birth dates. In children, the nutritional index weight for height ($W/H$) was computed using the package Anthro Version 1.01 (CDC/WHO, 1990).

### Food intake

Family food consumption was measured on two occasions during 1991, in January and in June, on a subsample of 35 compounds. Intake was measured using the precise weighing method: all foods were weighed before preparation and after cooking; leftovers from the finished meal were also weighed. Weighing was carried out on mechanical Terraillon balances (10 000 g ± 5 g) during four consecutive days. The six locally recruited observers also noted whether the foods were locally produced or bought on the market.

Energy and nutrient contents were estimated using a compilation of several tables of food composition (Toury *et al.*, 1967; FAO, 1968, 1972) suited for West African diets (Chevassus-Agnès, 1982). As meals are taken from a common dish, individual intakes could not be estimated. Instead, the total intake of each nutrient was divided by the sum of individual recommended dietary allowances of the consumers, giving a mean

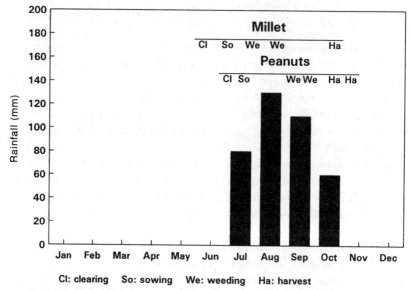

Cl: clearing    So: sowing    We: weeding    Ha: harvest

Fig. 13.3. Agricultural calendar in the Peanut Basin in 1991.

percentage of the recommended dietary allowance (% RDA) per nutrient and per food group (Chevassus-Agnès & Ndiaye, 1981). Energy needs were calculated using an estimate of the basal metabolic rate, BMR (which takes sex and age into account, FAO/WHO/UNU, 1985) together with estimates of energetic costs of physical activity.

### Activity patterns and total energy expenditure

The physical activity of one woman and one man per compound was monitored during the four days of food consumption survey, using the diary method (Edholm, 1981). Every 15 min. between 8.00 a.m. and 8.00 p.m. an observer noted the dominant activity during the previous 15 minutes. When the subject was absent from the compound, the activities were obtained by interviews when he/she returned. Activities were divided into 17 groups based on the level of energy expenditure.

Total energy expenditure (TEE) was estimated from the recorded activities using measurements of activity-specific energy costs measured in previous studies in Burkina Faso (Bleiberg *et al.*, 1980; Brun *et al.*, 1981) and Guatemala (Torun *et al.*, 1982). Night energy costs were set to $1.0 \times BMR$ for 8 hours and $1.2 \times BMR$ for 4 hours.

### The Niakhar study area

The Niakhar study area (the Peanut Basin, PB) is located about 150 km east of the capital city of Dakar. Its area of 230 km$^2$ has a population of 26 500 inhabitants. The study population belongs mostly (96%) to the Serere ethnic group. Main religions reported are Islam (72%) and Christianity (24%) but they have been introduced rather recently and animism is still very strong. Eighty-eight per cent of inhabitants above the age of eight years are farmers. Millet is grown for subsistence and peanuts are grown as a cash crop. Farming is totally dependent on rains, which occur between July and October. Mean annual rainfall, from measurements over a ten year period, is 414 mm (range: 227 to 560 mm, Niakhar Project, 1992).

The calendar of agricultural tasks is presented in Fig. 13.3. Clearing of the fields starts in May. Sowing of millet is done in June, just before the first rains, while peanuts need rain before sowing and are thus usually sown in the first half of July. The millet harvest takes place in late September–early October, while the peanut is harvested in October–November.

### Subjects and methods

#### Mortality

Since 1987 a continuous demographic surveillance system has ensured highly reliable information on deaths, births, migrations and age at weaning. Briefly, demographic field workers visit all compounds weekly and register those of the above mentioned events that have occurred since their previous visit. The database is updated weekly at the headquarters in Dakar. Accuracy of event registration is checked through a yearly demographic census.

#### Anthropometry

Immunization sessions are conducted monthly in each of the three dispensaries in the area. All infants residing in the study area are invited to the vaccination sessions at ages 2, 4, 6 and 9 months, to receive the standard EPI vaccines. Eligible children may be enrolled in vaccine trials in which a standard vaccine may be substituted with a test vaccine. Absent infants are called up again the following month. Coverage rate is about 80% per session.

During the vaccination session anthropometric measurements are made on all infants present using methods identical to those used in the SRV.

Since December 1989 anthropometric measurements have also been made on mothers. Height is measured with a Metabo stadiometer (precision: $\pm 0.1$ cm), headcloths being removed prior to measurement. They are also weighed, in the manner already described for the SRV. On average, 280 mothers (aged 15–50 years) and infants (aged 2–10 months) are measured each month.

Table 13.1. *Physical characteristics of adults in February 1991*

|  | Senegal River Valley | | Peanut Basin |
|---|---|---|---|
|  | Men ($n=91$) | Women ($n=162$) | Women ($n=265$) |
| Age (years) | 38.4±10.8 | 29.7±7.4*** | 29.6±7.1 |
| Height (cm) | 174.4±7.6 | 163.8±5.2*** | 161.4±6.0*** |
| Weight (kg) | 65.4±9.0 | 57.8±10.4*** | 56.3±7.0*** |
| MUAC (cm) | 27.4±2.4 | 25.7±3.3*** | 26.7±2.6*** |
| BMI (kg/m²) | 21.5±2.8 | 21.5±3.3 | 21.6±2.2 |

Values are mean±SD.
***$p<0.001$.
MUAC: mean left upper arm circumference, BMI: body mass index.

### Morbidity

Since 1987 a morbidity surveillance system has covered all resident infants aged 2–12 months. During weekly interviews with the mothers, field workers record the dates of onset and end of 46 different symptoms. Only the two most prevalent symptoms, fever (34.2%) and diarrhoea (19.1%), are considered here.

### Statistical analyses

Analysis of variance is used for general comparison of means. *T*-tests are used for comparison of means, two by two. In the case of multiple comparisons, the significance level is adjusted accordingly using Bonferroni's Inequality. The Kruskall–Wallis non parametric test is used in the analysis of food intake. All analyses are carried out using the BMDP statistical package (University of California, 1990).

## Results

### Anthropometry

### Adults

In the Senegal River Valley men were significantly older than women (38.4 years versus 29.7 years, $p<0.001$; Table 13.1), but there were no sex differences in the mean body mass index (21.5 kg/m² in both sexes).

Body dimensions differed significantly between women in the two settings. The Toucouleur women were taller (163.8 cm versus 161.4 cm, $p<0.001$), while the Serere women had thicker arms (26.7 cm versus 25.7 cm, $p<0.001$). Mean body mass index did not differ between Toucouleur and Serere women (21.5 kg/m² versus 21.6 kg/m²).

Prevalences of malnutrition among adults in February 1991 were low in both settings (Table 13.2). Using the cutoff points proposed by James *et al.*

Table 13.2. *Distribution of body mass index in adults (percentages)*

| | Senegal River Valley | | Peanut Basin |
| --- | --- | --- | --- |
| BMI (kg/m²) | Men ($n=91$) | Women ($n=162$) | Women ($n=265$) |
| <16.0 | 1.1 | 3.1 | 0.0 |
| 16.0–16.9 | 1.1 | 1.2 | 0.0 |
| 17.0–18.4 | 4.4 | 10.5 | 5.3 |
| ≥18.5 | 93.4 | 85.2 | 94.7 |

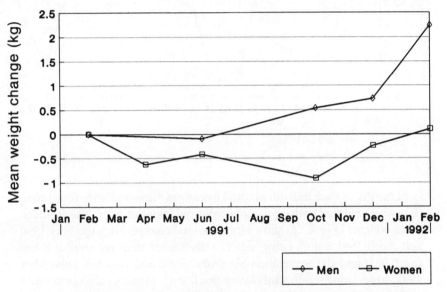

Fig. 13.4. Mean change in individual weight of adults in the Senegal River Valley.

(1988), at most 6.6% of the men and 14.8% of the women in the SRV are classified as malnourished, against 5.3% of the women in the PB.

The means of individual weight changes of adults in the SRV between successive measurements across the year are given in Fig. 13.4. Patterns differ between the sexes. For men, weight increased significantly between June and October ($+0.63$ kg, $p<0.05$) and between December and February ($+1.5$ kg, $p<0.001$), while women lost significant amounts of weight between June and October ($-0.49$ kg, $p<0.05$). Thereafter they regained the weight lost: between October and December ($+0.67$ kg, $p<0.01$) and between December and February ($+0.35$ kg, $p<0.05$).

Maximal weight change in 1991 is estimated to be 1.9 kg for women and 0.7 kg for men. A positive trend in the nutritional status of men is clear with the February 1992 weight being about 2 kg above the February 1991 weight.

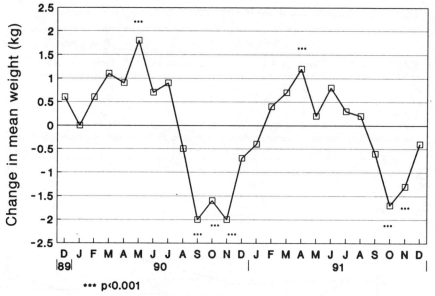

Fig. 13.5. Seasonal change in mean weight of women in the Peanut Basin.

In the PB, as each individual could have been measured up to four times, data were analyzed both cross-sectionally and longitudinally. The mothers' mean weights (Fig. 13.5) show large seasonal changes with the May 1990 and April 1991 values being significantly higher than the overall mean ($p < 0.001$) and the September–November 1990 and October–November 1991 values being significantly lower ($p < 0.001$). Maximal change in mean weight was 3.8 kg in 1990 and 2.9 kg in 1991.

Mean individual weight changes for women weighed both in April–May and September–November were 3.5 kg in 1990 (95% confidence, interval: 3.0–4.1, $n = 87$) and 2.5 kg in 1991 (95% confidence, interval: 2.0–3.0, $n = 145$).

### Infants and children

Seasonal variation in nutritional status in infants showed similar patterns in the two study areas. In the SRV, mean weight for height varied significantly between seasons ($p < 0.01$), with the June and October means being lower than the general mean ($p < 0.05$, Fig. 13.6). Between October and December the mean increased sharply from $-0.40$ $z$-scores to $+0.28$ $z$-scores and then it declined again between December 1991 and February 1992. The seasonal pattern was similar for children aged 12–23 months while the weight for height of 24–59-month-old children showed no significant variation.

The PB younger infants (2–5 months) exhibited small and inconstant

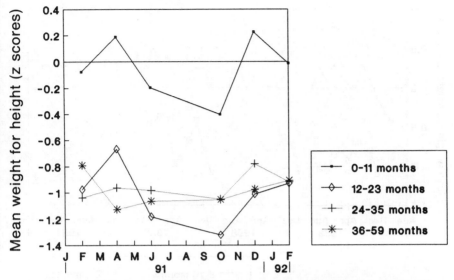

Fig. 13.6. Seasonal change in weight for height of preschool children in the Senegal River Valley.

declines in *W/H* at the end of the rainy season, while the maximal difference in *W/H* of older infants (9–10 months) was consistently large (range: 0.6–0.8 *z*-scores, Fig. 13.7). The decline was abrupt from July to September–October, and the increase was rapid from October–November to December.

### Food intake in the Senegal River Valley

Energy intakes measured at the household level were globally sufficient to cover dietary allowances. Mean energy intake was significantly higher in January than in June (10.8 MJ/d (2591 kcal/d) versus 9.8 MJ/d (2353 kcal/d), $p < 0.05$).

In January, just after the rice harvest, rice consumption covered about 50% of the total energy intake. In June, significantly more foods were bought on the market than in January.

### Total energy expenditure of adults in the Senegal River Valley

Total energy expenditure was higher in January than in June, both for men (11.1 MJ/d (2660 kcal/d) versus 10.1 MJ/d (2415 kcal/d), $p < 0.01$) and for women (11.5 MJ/d (2751 kcal/d) versus 10.0 MJ/d (2394 kcal/d), $p < 0.01$).

When expressed in multiples of BMR, women had higher energy expenditure than men in both seasons ($2.0 \times$ BMR in January and $1.8 \times$ BMR in June versus $1.6 \times$ BMR in January and $1.5 \times$ BMR in June). Total energy expenditure divided by estimated individual energy intake

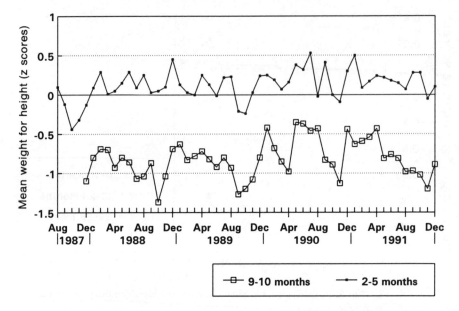

Fig. 13.7. Seasonal change in weight for height of infants in the Peanut Basin.

was constant for men (110% in January and 106% in June), and not significantly lower in January than in June for women (105% versus 112%).

### Activity patterns in the Senegal River Valley

Women devoted an important part of their day to housework, mainly food preparation. Housework was either moderate or heavy in intensity. Moderate housework took up 29% of the day in January and 24% in June. Heavy activities such as pounding and drawing water took up 13% of the day in January and 12% in June (Table 13.3). Agricultural tasks involved 10% of the day in January and 6% in June, while the women were sitting, either active or inactive, for 25% of the day in January and 37% in June.

Men devoted 20% of the day to agriculture in January and 15% in June, while they were sitting, mostly inactive, during 32% of the day in both seasons. They took part in significantly more social activities such as games and religious ceremonies in June than in January (15% versus 7%, $p < 0.05$). They also remained lying more (10% versus 5%, $p < 0.01$) and moved around less in June than in January (9.2% versus 15.8%, $p < 0.05$).

### Morbidity during infancy in the Peanut Basin

No clear seasonal pattern was observed in the number of diarrhoea cases in children aged 2–12 months between 1988 and 1990 (Fig. 13.8).

Table 13.3. *Physical activity of adults (percentage of the day)*

| | Women ($n=40$) | | Men ($n=30$) | |
|---|---|---|---|---|
| | January | June | January | June |
| Moderate housework | 28.7 | 24.0 | 3.9 | 3.5 |
| Heavy housework | 12.9 | 11.5 | 1.5 | 2.1 |
| Agricultural tasks | 10.0 | 5.6* | 19.9 | 15.3 |
| Other heavy tasks | 5.3 | 4.4 | 14.6 | 12.6 |
| Moving around | 9.7 | 5.9 | 15.8 | 9.2* |
| Social activities | 4.5 | 4.7 | 7.0 | 14.9* |
| Sitting | 24.9 | 37.0*** | 32.3 | 32.3 |
| Lying | 4.0 | 6.8* | 4.6 | 10.0** |

Differences between seasons: *$p<0.05$, **$p<0.01$, ***$p<0.001$.

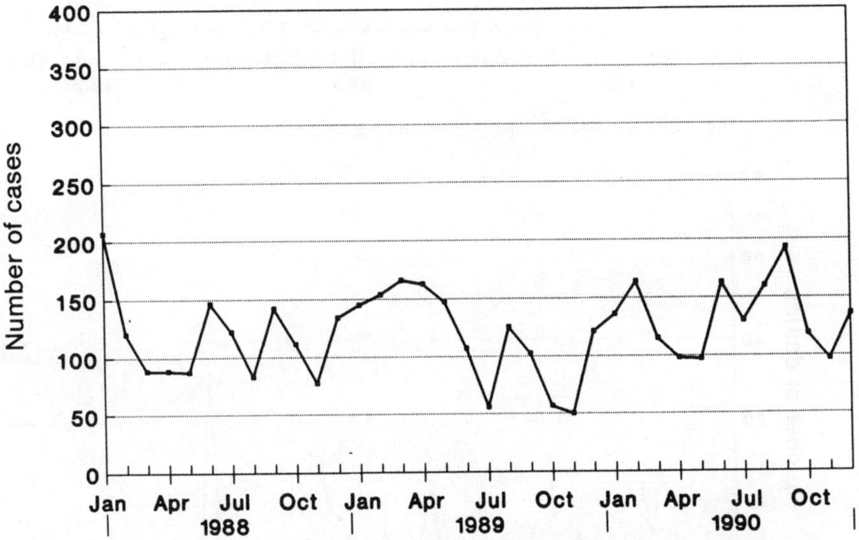

Fig. 13.8. Seasonal change in diarrhoea morbidity among infants in the Peanut Basin.

In contrast, the number of fever cases was two- to three-fold higher in September–October 1988 and 1989 than in July (Fig. 13.9). In 1990, the peak at the end of the rainy season was lower, although still marked.

### Mortality during early childhood in the Peanut Basin

Mortality during early childhood was highly related to season. The number of fatality cases in September–November was five- to ten-fold higher than in February–April, both for infants (Fig. 13.10) and for one- to four-year-old children (Fig. 13.11). In both age groups the yearly minimal value was about four deaths per month while the yearly maximal value varied

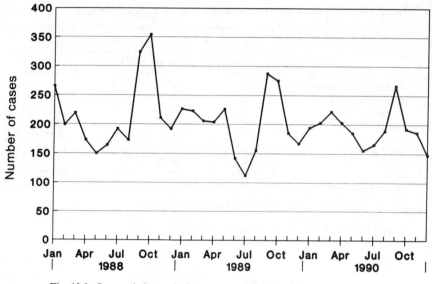

Fig. 13.9. Seasonal change in fever among infants in the Peanut Basin.

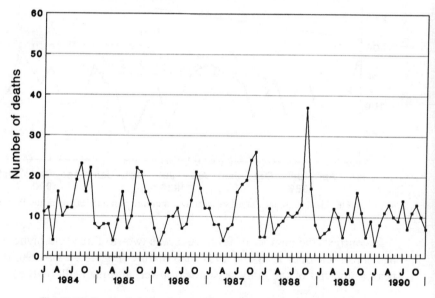

Fig. 13.10. Seasonal change in mortality of infants in the Peanut Basin.

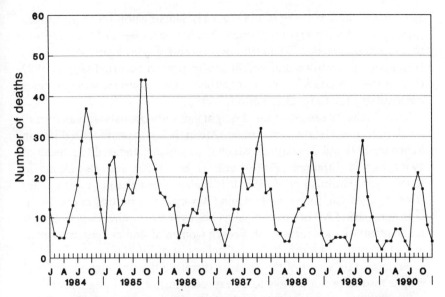

Fig. 13.11. Seasonal change in mortality of children aged 1–4 years in the Peanut Basin.

between 20 and 40 deaths per month. However, the peak during the end of the rainy season was not observed in the last two years (1989–90) for infants.

### Discussion

The magnitude of mean seasonal weight loss experienced by women in the SRV (1.9 kg) fits in the range reported for developing countries (Ferro-Luzzi *et al.*, 1987). Mean weight loss was lower for men (0.7 kg) than for women, while most other studies report higher weight losses for men. The most striking phenomenon in the nutritional status of men is the positive annual trend. This result is consistent with the differences observed in total energy expenditure, which is higher for women than for men in both seasons (1.8–2.0 × BMR versus 1.5–1.6 × BMR). Unfortunately, in the absence of data on seasonality and nutritional status in this area before the onset of the agricultural project, the impact of the project on seasonality cannot be assessed.

The weight loss experienced by women in the PB (3.5 kg in 1990 and 2.5 kg in 1991) is one of the highest ever reported in the literature (Ferro-Luzzi *et al.*, 1987). In 1981–2, Rosetta (1986) studied seasonal anthropometric variation in Serere adults living in an area located only 25 km from the Niakhar study area. Men aged 18–54 years (mean age: 37.7 years) had a mean loss of 1.7 kg between March and August–September in

1981 ($n = 53$) and of 1.8 kg in 1982 ($n = 43$). Nursing non-pregnant women (mean age: 27.3 years) comparable to the Serere mothers of the infants in the present study lost 2.5 kg in 1981 ($n = 19$) and only 0.1 kg in 1982 ($n = 16$). It is, however, possible that weight loss in women occurred later in 1982 than in 1981 (in late September or October), but no weight measurements are available for those later periods.

Thus, in the PB severe seasonal weight variation coexists with an overall good nutritional status in women. Mean body mass index (BMI) in February 1991 was 21.6 kg/m$^2$, which is very high compared to a mean of 19.0 kg/m$^2$ in Ethiopia (Ferro-Luzzi *et al.*, 1990) and 18.5 kg/m$^2$ in Bangladesh (Huffman *et al.*, 1985). It is also higher than in most studies from West Africa, for example Benin (20.4 kg/m$^2$, Schultink *et al.*, 1990).

In fact, mean BMI of Toucouleur as well as Serere mothers is very similar to the reference data for well-off French women of similar ages (median at age 30 years: 21.1 kg/m$^2$, Rolland-Cachera *et al.*, 1991).

An important result of this study is therefore that the highest seasonal weight losses are not always to be expected in low BMI populations.

In addition, the seasonal weight losses reported here are probably slightly underestimated. In the SRV, subjects were only weighed every two to four months while, as the data from the PB show, weight variation occurs rather rapidly and the months of minima and maxima often change from one year to another, probably according to variations in the agricultural calendar. In the PB, mean weight was known for each month, but as rather few women had been weighed both in the month of maximal weight and in the month of minimal weight, the mean individual weight loss was calculated from a two month period in Spring (April–May) and a three month period in Autumn (September–November).

Seasonal variation in nutritional status of women was more important in the PB than in the SRV. A possible explanation for this difference is the rather homogeneous distribution of agricultural tasks throughout the year in the SRV, perhaps associated with better possibilities for purchasing foods in periods of food shortage. However, women lose weight during the rainy season in the SRV. Unfortunately, neither physical activity surveys nor food intake surveys were carried out in September–October, therefore the relative importance of these two factors could not be assessed.

No food consumption surveys are available from the PB area. Rosetta (1988) reports the results of family food consumption surveys in a similar area in two seasons in 1981: during the middle of the dry season (February–March) and at the end of the rainy season (August–September). The methodology was identical to that used here, and the study found no seasonal variation in energy intake at the household level (9.1 MJ/d (2182 kcal/d) during the dry season versus 8.8 MJ/d (2103 kcal/d) during the rainy season). When food consumption is measured at the household level, intakes of subgroups, such as women, cannot be assessed. In particular, it is impossible to know whether the energy allocation among the members of the food group is performed in the same way in different seasons.

No data on physical activity were available from this study, but it is well documented from similar areas that September is the peak season for agricultural activities, involving both men and women (Prentice *et al.*, 1981; Bleiberg *et al.*, 1980; Brun *et al.*, 1981). In farmers from Burkina Faso, energy expenditure of women was 9.7 MJ/d (2320 kcal/d) in the dry season versus 12.1 MJ/d (2890 kcal/d) in the rainy season (Bleiberg *et al.*, 1980).

In both areas studied here, variation in the nutritional status of infants is similar to that in the nutritional status of their mothers. Is there a causal relationship, and if so, through what mechanisms does it operate? One hypothesis is that breast milk production may decline when women lose weight beyond a certain level. However, in the PB, younger infants, who are almost exclusively breastfed, suffered much less from seasonal weight losses than older infants.

Another hypothesis is that the intake of solids in infants and very young children may be reduced during periods of intensive field work because mothers have less time for child-care.

Morbidity (especially diarrhoea and fever) is an important determinant of $W/H$ in young children (Black *et al.*, 1984; Rowland *et al.*, 1977). However, no increase in diarrhoea was seen during the rainy season in the PB. The absence of seasonal variation is unlikely to be explained by deficiencies in data quality, since the number of fever cases, collected a similar way, revealed significant seasonal variation. Black *et al.* (1982) also found little seasonal variation in the prevalence of diarrhoea.

The important increase in the number of fever cases at the end of the rainy season is probably a result of the seasonal increase in incidence of malaria.

The high mortality rates at the end of the rainy season in the PB are probably due in part to malaria, and in part to malnutrition. Even if diarrhoea prevalence does not increase at the end of the rainy season, diarrhoea fatality rates may be higher because of the high prevalence of acute malnutrition. The relative importance of these factors on the seasonal mortality peaks will be assessed in further studies through the analysis of cause-specific mortality rates.

### Acknowledgements

The multidisciplinary health research programme *Eau et Santé dans les Contextes du Développement* conducted by ORSTOM in the Senegal River

Valley was supported by the French *Ministère de la Recherche et de la Technologie*.

The demographic census was carried out by Pascal Handschumacher. Marie Sy-Ndiaye (ORANA) and Daouda Ndiaye supervized the food consumption and activity surveys, while Pape Niokhor Diouf took the anthropometric measurements and Oumar Sall served as interpreter. Simon Chevassus-Agnès provided the software for the analysis of food consumption data.

Between 1987 and 1991 the Niakhar Study Area was supported in part by the Task Force for Child Survival, Atlanta, the World Health Organization, Geneva and Pasteur-Mérieux, Paris. Until August 1989, Michel Garenne coordinated research activities and managed the demographic surveillance system in the area. Anthropometric measurements of infants and mothers were taken under supervision of Bassirou Fall. The field staff in Niakhar collected the demographic and morbidity data, while the staff in Dakar handled data entry and verification.

Study subjects in the Senegal River Valley and in Niakhar collaborated with kindness and patience.

### References

Bénéfice, E. & Chevassus-Agnès, S. (1985). Variations anthropométriques saisonnières des adultes appartenant à deux populations différentes de l'Afrique de l'Ouest. *Révue d'Epidémiologie et de Santé Publique*, **33**, 150–60.

Bénéfice, E., Chevassus-Agnès, S. & Barral, H. (1984). Nutritional situation and seasonal variations for pastoralist populations of the Sahel (Senegalese Ferlo). *Ecology of Food and Nutrition*, **14**, 229–47.

Black, R.E., Brown, K.H., Becker, S. & Yunus, M. (1982). Longitudinal studies of infectious diseases and physical growth of children in rural Bangladesh. I. Patterns of morbidity. *American Journal of Epidemiology*, **115**, 305–14.

Black, R.E., Brown, K.H. & Becker, S. (1984). Effects of diarrhea associated with specific enteropathogens on the growth of children in rural Bangladesh. *Pediatrics*, **73**, 799–805.

Bleiberg, F., Brun, T. & Goihman, S. (1980). Duration of activities and energy expenditure of female farmers in dry and rainy season in Upper Volta. *British Journal of Nutrition*, **43**, 71–82.

Brun, T., Bleiberg, F. & Goihman, S. (1981). Energy expenditure of male farmers in dry and rainy season in Upper Volta. *British Journal of Nutrition*, **45**, 67–75.

CDC/WHO (1990). *Anthro Version 1.01*. Atlanta: CDC.

Chevassus-Agnès, S. (1982). *Approche des besoins dans le logiciel de traitement de l'ORANA pour les enquêtes de consommation par groupes alimentaires*. Dakar: ORANA.

Chevassus-Agnès, S. & Ndiaye, A.M. (1981). Enquêtes de consommation alimentaires de l'ORANA de 1977 à 1979: méthodologie et résultats. In *Etat nutritionnel de la population rurale du Sahel: Rapport d'un groupe de travail*, pp. 57–66. Ottawa: CRDI.

Edholm, O. (1981). Habitual physical activity and energy expenditure. In *Practical*

*Human Biology*, ed. J. Weiner & J. Lourie, pp. 227–39. London: Academic Press.

FAO (1968). *Tables de composition alimentaire pour l'Afrique.* Rome: Food and Agriculture Organization.

FAO (1972). *Tables de composition alimentaire pour l'Asie du Sud-Est.* Rome: Food and Agriculture Organization.

FAO/WHO/UNU (1985). *Energy and Protein Requirements.* FAO/WHO/UNU Expert Consultation. Geneva: WHO.

Ferro-Luzzi, A., Pastore, G. & Sette, S. (1987). Seasonality in energy metabolism. In *Chronic Energy Deficiency: Consequences and Related Issues*, ed. B. Schürch & N.S. Scrimshaw, pp. 37–58. Lausanne: IDECG.

Ferro-Luzzi, A., Scaccini, C., Taffese, S., Aberra, B. & Demeke, T. (1990). Seasonal energy deficiency in Ethiopian rural women. *European Journal of Clinical Nutrition*, **44** (Suppl. 1), 7–18.

Gessain, M. (1978). Poids individuels saisonniers chez les Bassaris du Sénégal Oriental. *Bulletin et Mémoires de la Société d'Anthropologie de Paris*, **5**, 149–55.

Huffman, S.L., Wolff, M. & Lowell, S. (1985). Nutrition and fertility in Bangladesh: nutritional status of nonpregnant women. *American Journal of Clinical Nutrition*, **42**, 725–38.

James, W.P.T., Ferro-Luzzi, A. & Waterlow, J.C. (1988). Definition of chronic energy deficiency in adults. *European Journal of Clinical Nutrition*, **44**, 969–81.

Loutan, L. & Lamotte, J.-M. (1984). Seasonal variations in nutrition among a group of nomadic pastoralists in Niger. *Lancet*, **ii**, 947–9.

Niakhar Project (1992). *Population et santé à Niakhar. Niveaux et tendances des principaux indicateurs démographiques et épidémiologiques de la zone d'étude. 1984–1991.* Dakar: ORSTOM.

Prentice, A.M., Whitehead, R.G., Roberts, S.A. & Paul, A.A. (1981). Long-term energy balance in child-bearing Gambian women. *American Journal of Clinical Nutrition*, **34**, 2790–9.

Rolland-Cachera, M.-F., Cole, T.J., Sempé, M., Tichet, J., Rossignol, C. & Charraud, A. (1991). Body mass index variations: centiles from birth to 87 years. *European Journal of Clinical Nutrition*, **45**, 13–21.

Rosetta, L. (1986). Sex differences in seasonal variations of the nutritional status in Serere adults in Senegal. *Ecology of Food and Nutrition*, **18**, 231–44.

Rosetta, L. (1988). Seasonal variations in food consumption by Serere families in Senegal. *Ecology of Food and Nutrition*, **20**, 275–86.

Rowland, M.G.M., Cole, T.J. & Whitehead, R.G. (1977). A quantitive study into the role of infection in determining nutritional status in Gambian village children. *British Journal of Nutrition*, **37**, 441–50.

Schultink, W.J., Klaver, W., Van Wijk, H., Van Raaij, J.M.A. & Hautvast, J.G.A.J. (1990). Body weight changes and basal metabolic rates of rural Beninese women during seasons with different energy intakes. *European Journal of Clinical Nutrition*, **44** (Suppl. 1), 31–40.

Torun, B., McGuire, J. & Mendoza, R. (1982). Energy cost of activities and tasks of women from a rural region of Guatemala. *Nutrition Research*, **2**, 127–36.

Toury, R., Giorgi, R. & Favier, J.-C. (1967). Aliments de l'Ouest africain, tables de composition. *Annales de Nutrition et d'Alimentation*, **21**, 73–127.

University of California (1990). *BMDP statistical Software.* Berkeley: University of California.

# 14 Culture, seasons and stress in two traditional African cultures (Massa and Mussey)

IGOR DE GARINE

Over the last two decades, the influence of seasonality has appeared as a central theme in human ecology. In most cases the focus has been on objective aspects: measurable biological variations such as mortality, morbidity, birth rate, growth rate and state of nutrition in relation to seasonal environmental changes (e.g. Chambers *et al.*, 1981; Harrison, 1988, p. 27; Foeken, 1990). In human societies, these aspects are influenced by cultural choices, which are more or less attuned to optimal biological fitness. Cultural factors which have a bearing on human biology and are influenced by seasonality include lifestyle, type of production system, choices of technology, division of labour, social organisation, time schedule, habitat choice, food consumption, and aspects of material culture. However, humans are endowed with symbolic thinking, and the non-material, psychocultural (Folkman & Lazarus, 1988) and cognitive (Laughlin & Brady, 1978, pp. 1, 13) aspects of seasonality are also important. Butzer (1982, p. 255) writes 'identification of environmental components, resource opportunities . . . are all intimately related to group perception'. Through culture, the material environment is grasped as a cognised environment, an 'intentional world' (Shweder, 1990, p. 25) endowed with meaning and calling for actions 'on tangible as well as on non-tangible realities' (Augé, 1986, p. 81), each society having its own local symbolic system (Stigler *et al.*, 1990, p. VIII).

Perception of events and coping behaviour have emotional components; they are appraised as positive/good or negative/bad according to personal and cultural views. Geertz wrote 'Not only ideas but emotions too are cultural artefacts in Man . . . the kinds of culture symbols that serve the intellective and affective side of human mentality tend to differ . . . but the contrast should not be drawn too sharply' (Geertz, 1973, p. 81). D'Andrade (1984, p. 101) recognises that even 'kin terms [have] a core of culturally constructed, highly affective and directive elements as well as a representational aspect'. Among the Massa, the term *baknarda* (stark famine) is laden

184

with such emotion that the elderly, who have experienced it, do not dare to utter the word.

### 'Stress'

In relation to emotions, the terms 'stress' and 'well-being' are constantly and loosely used in the general and ecological literature, but specific biological meanings are not necessarily implied. The concept of stress is simply 'any stimulus, as fear or pain, that disturbs or interferes with the normal physiological equilibrium of an organism; or physical, mental or emotional strain or tension'. (Random House Dictionary, 1979). The opposite concept is 'well-being', a vague notion 'encompassing physical health, satisfaction and happiness' (Colby, 1987, p. 879). The term 'stress', as in general usage, may be the same as that defined by biologists (Selye, 1956; Mims, 1981), and which may have negative consequences when regulating mechanisms can no longer cope (Cathebras & Rousset, 1988). It may be associated with hypertension and cardiovascular disease (Boyden, 1970, p. 199; Mims, 1981, p. 178), disturbance of digestive function (Kelly, 1966; Wolf, 1965), growth retardation (Widdowson, 1951), psychosomatic disease (Parkes, 1972) and susceptibility to infection (Andreoli *et al.*, 1989, p. 36; Schleifer *et al.*, 1984; James *et al.*, 1989, p. 288). It may also predispose individuals to mental illness (Simons & Hughes, 1985). In this article the term 'stress' is used to refer to negative or negatively-appraised events.

In any group or individual, physical and mental health, biological fitness and psychosocial well-being do not necessarily coincide, and periods and events that are judged to be negative by outside observers are not necessarily perceived as such by those undergoing them. For example, fasting to the verge of starvation would hardly appear to the Western nutritionist as a positive experience, on biological grounds. However, it is considered to be a highly rewarding act by a Hindu Brahmin, for psychocultural reasons. Conversely, periods and events which would not be considered as stressful according to Western, 'rational' views may appear as deleterious to those involved. They can result in high emotional arousal, as shown, for instance, by Pagezy (1986, 1988) when dealing with the effects of meat hunger among the Oto and Twa of Lake Tumba in Zaire. In this article, an attempt will be made to describe the objective and subjective effects of seasonality on two African societies, the Massa and the Mussey of Cameroon.

### Methods

This account is based on detailed ethnographic work and study of the Massa and Mussey food systems. Quantitative food consumption surveys and nutritional anthropometry covered three annual cycles, in 1976–7, 1980–1 and 1984–5. Ethnographic data are based on six years of participant observation during the periods 1958–65, 1980–5 and 1988–92. In particular, key informants' records, interviews and open-ended questioning on daily life and ritual life allowed most life events to be studied in relation to season. The lexicon and oral literature of these groups provide additional evidence relating to the perception of seasons.

## Material and objective aspects of seasonality

### Seasonal ecological constraints

The Massa and the Mussey are two savannah populations of Northern Cameroon and Chad, each totalling about 250 000 inhabitants. They are spread along flood plains of the Logone River 230 km south of N'jamena, the capital of Chad, where they practise mixed economies (de Garine, 1980, p. 44). They both experience preharvest food shortages similar to those endured by people elsewhere in tropical Africa (e.g. Bénéfice & Chevassus-Agnès, 1985; Teokul *et al.*, 1986; Rosetta, 1988). Problems occur at the time when food shortages coincide with the heaviest energy expenditure needs (e.g. Fox, 1953; Schultink, 1991).

A single dry season operates from November to May, and one rainy season from June to October. However, five sub-seasons can be distinguished: 1. dry and cold, between mid-November and mid-February; 2. dry and hot, between mid-February and mid-May; 3. damp and warm, between mid-May to mid-June; 4. rainy and temperate, between mid-June and mid-September; and 5. damp and progressively hotter, between mid-September and mid-November.

The rains are very irregular, the yearly average being around 800 mm. The amount varies sharply from year to year (rainfall in Géré in 1981 was 919 mm, in 1985 it was 547 mm) and from place to place (in 1988 rainfall in Gobo was 1146 mm, while in Géré (40 km west of Gobo) it was 737 mm). The monthly distribution of rainfall fluctuates, and droughts following the planting periods often kill planted seedlings, often resulting in crop failure. Continental winds (Harmattan) blow from the north-east during the dry season, monsoon winds from the south-west during the rainy season. The hottest month of the year is March, the coldest, January. Temperature differences between daytime maxima and nighttime minima are about 8 °C in August, but about 22 °C in March. Direct climatic stresses are: 1. heat, in March and April; 2. cold, due to large differences between day and nighttime temperature, and due to cold winds between November and January; and 3. dust, brought by the Harmattan wind between November and February.

*Disease*

Information on disease patterns comes from the inventory of consultations of mostly Mussey people at the local dispensary at Gobo. Some 6800 individuals were seen between December 1975 and November 1976. For the Massa, similar information was obtained by the retrospective questioning of 115 adults at two times of year: June (dry season); and October (wet season) (Koppert, 1981, pp. 129, 134).

Three distinct periods of peak infection are evident. The first is during the rainy season, when malaria, respiratory diseases, skin ailments, diarrhoea, otitis and anaemia are present and the workload is heavy. It should be noted that diarrhoea and malaria are present at some level throughout most of the year. The second is during the dry, cold season when it is cold, windy and dusty; ailments at high incidence include rheumatism, venereal diseases, respiratory diseases, pneumonia, children's illnesses such as measles and chicken pox, and burns. In addition, epidemics of meningitis are more likely to occur at this time of year than at any other. Respiratory diseases may be related to generally low temperatures and to sharp swings in temperature between daytime and nighttime; growing dryness may desiccate the mucosa and diminish natural defence to bacteria brought by the dusty Harmattan wind (Beauvilain, 1989, p. 193). Epidemics of disease have taken a heavy toll on these populations in the past, and epidemics of some diseases still take place (Beauvilain, *op. cit.*, pp. 160–90). Epidemics of the following diseases have been documented: 1. Smallpox in 1924, 1931, 1950 and 1953; 2. Cerebrospinal meningitis in 1943, 1945, 1946, 1951, 1975, 1976 and 1990; and 3. Cholera in 1971, 1984, 1985, 1990 and 1992. The third seasonal peak in infection comes during the dry, hot season when schistosomiasis and infections of the urinary system occur, as do a number of children's ailments and dehydration.

As in other Northern Cameroonian groups (Gubry, 1990, p. 6), child mortality is at its peak at the end of the dry, hot season and at the end of the rainy season; for the elderly, mortality is at its highest at the end of the rainy season and during the dry, cold season.

*Nutrition*

The nutritional aspects and material coping systems of the Massa and Mussey have been documented elsewhere (de Garine and Koppert, 1988) and will not be considered at any length here. Their diets are low in vitamin C during the dry season, while energy intakes are low during the rains. The daily energy intake, averaged across the year, is about 10.5 MJ (2500 kcal), reaching about 12.5 MJ (3000 kcal) during the crop season, and 9.6 MJ (2300 kcal) in the rainy season, during good years (e.g. 1976). In bad years, energy intakes during the rainy season are much lower, when typically only one meal per day is taken and days without food are common. There are also marked declines in body mass as a result of negative energy balance during the rainy season (Koppert, 1981, p. 139; de Garine & Koppert, 1988, p. 229). The food shortage lasts between 4 and 6 weeks and usually takes place between mid-July and the beginning of September, resulting in average losses of weight of about 2 kg for adults of both sexes in good and normal years, and about 5 kg for women and 7 kg for men in bad years. The values for normal years are similar to those reported for the Gambia (Lawrence *et al.*, 1985) and Burkina Faso (Brun *et al.*, 1981, p. 67).

During the rainy season, 19% of the children have weight for height below 80% of the National Center for Health Statistics' standard, with 38% being below 85% of standard. In the dry season, 8% of the children have weight for height below 80% of standard, and 23% below 85% of standard (Froment & Koppert, 1989). There are two peaks in birthweight. The first is in June, at the end of the dry season; and the second is in November, after the harvest. The lowest birthweights are in August, during the rainy season (Froment & Koppert, op. cit.).

Each year, the Massa and Mussey experience food shortages during the rainy season. These cannot be averted by buying food because money is not available at this time. The storage strategy of putting aside cereals for the field tilling period goes some way toward alleviating possible food shortages in the first part of the rainy season (de Garine & Koppert, 1988, p. 235); and hunger is at its worst during the second part of the rains, when food stores run out (mid-July to September). This is usually a 'manageable' period of food restriction, although in some years it may degenerate into famine, when drought, insect infestation and, in the past, raiding, combine to make life more difficult than usual. Under these circumstances, some people die. Beauvilain (op. cit., p. 112–27) records famines, with local variations in intensity, in 1880, 1890–3, 1912–14, 1925–6, 1930, 1939, 1955, 1961, 1972–3 and 1979–80. In 1985, a situation close to famine was experienced, after two consecutive years of low rainfall (annual rainfall, 1984: 547 mm; 1985: 647 mm) and partial crop failure.

The staple food, sorghum, is consumed by the Massa and the Mussey year round, but other parts of their diet vary with season. This is particularly true of foods that are low in energy but high in micronutrients. Protein sources include: fish, which is available all year round except during the time of heavy flooding (mid-July to mid-September); and milk, which is available (mostly to men) between June and January.

From December to May, the Massa diet consists predominantly of fish. During the rainy season, and especially during the shortage period, both groups rely on green leaves and wild tubers. In good years, the alternation

|  |  | DEC | JAN | FEB | MAR | APR | MAY | JUN | JUL | AUG | SEP | OCT | NOV |
|---|---|---|---|---|---|---|---|---|---|---|---|---|---|
| Heavy work | Rain sorghum |  |  |  |  |  |  |  |  |  |  |  |  |
|  | Dry season sorghum |  |  |  |  |  |  |  |  |  |  |  |  |
|  | Cotton |  |  |  |  |  |  |  |  |  |  |  |  |
|  | Harvest transportation |  |  |  |  |  |  |  |  |  |  |  |  |
| Light work | House-building |  |  |  |  |  |  |  |  |  |  |  |  |
|  | Basket making |  |  |  |  |  |  |  |  |  |  |  |  |
| Prestigious occupations | Hunting |  |  |  |  |  |  |  |  |  |  |  |  |
|  | Fishing |  |  |  |  |  |  |  |  |  |  |  |  |

Fig. 14.1. Seasonal distribution of workload for the Massa and Mussey.

| | OCT | NOV | DEC | JAN | FEB | MAR | APR | MAY | JUN | JUL | AUG | SEP |
|---|---|---|---|---|---|---|---|---|---|---|---|---|
| Season | DRY | | | | | | | | RAINY | | | |
| Temperature | Warm to cool and dry | | Cold - Windy Dusty mist | | Very hot and dry | | | | Damp and cooler | Rainy and cool | | Damp warm |
| Disease | Respiratory, Venereal diseases - Burns | | | | Strong fevers | | | | Malaria - Diarrhoea | | | |
| | Epidemic diseases - Meningitis - Cholera - Measles - Rabies | | | | | | | | Snake bites | | | |
| Energy Expenditure | MEDIUM Harvesting | | LOW Hunting - Fishing - Building | | MEDIUM | | HIGH Tilling - Planting - Weeding | | VERY HIGH | | HIGH Harvesting | |
| Grain Stores | ABUNDANCE | | RESTRICTION (by choice) | | | | RAINS GRANARY | | SHORTAGE (unavoidable) | | FIRST CROP | |
| Money | AVAILABLE (Cash crop) SOCIAL EXPENSES | | | | LACKING TAXES - BUYING FOOD | | | | | | | |
| Sorghum and Millet | | | | | | | | | | | | |
| Pulses | | | | | | | | | | | | |
| Bush Cereals/Tubers | | | | | | | | | | | | |
| Greens and Fruits | | | | | | | | | | | | |
| Milk | | | | | | | | | | | | |
| Fish | | | | | | | | | | | | |

Fig. 14.2. Food and well-being calendar, Massa and Mussey.

of these foods gives some variety to an otherwise monotonous diet. There may be some health benefits from this pattern of consumption; Etkin & Ross (1982) have contended that in northern Nigeria the shift from a cereal diet to one where green leaves are abundant might help to fight helminth infection.

### Workload

Fig. 14.1 shows the seasonal distribution of workload for the Massa and Mussey. As in other populations living in the tropical African savannah, the heaviest workload occurs at the end of the dry season when fields have to be cleared, ridged, planted and weeded. Total energy expenditures for both populations combined are, in multiples of resting metabolic rate, 1.63 (men) and 1.77 (women) (Pasquet *et al.*, in press). Two other busy periods are February–June, when carrying drinking water is the heaviest task, and September–October, when the main crops are harvested and carried into storage (Fig. 14.2).

## Subjective, non-material, aspects of seasonality

### Concepts of time and season

The Massa and Mussey year has twelve lunar months, each of 28 days. Months are counted from the appearance of the first quarter after the new moon. They are distinguished by objective criteria such as the position of stars, meteorological events (especially wind direction), botanical observations (e.g. Grivetti, 1981, p. 516), animals' behaviour, and increased numbers of certain types of insect. Bird migration, nesting and pregnancy in wild animals are also markers of specific times of year. During the reproductive period, birds and animals disappear; like humans, they are believed to be confined because of the rains. During the wet season (*dolla*), everything is 'shut in'. It is a season when mysterious processes such as the reproduction of animals and plants take place; a respectful mood should be adopted to avoid disturbing the highly receptive supernatural powers at work.

### Supernatural powers

Seasonal variations are ultimately traced to supernatural powers directly or through human intermediaries. Earth priests, rain chiefs, possessed individuals, diviners, elder members of a lineage or simply household heads are helpful specialists (e.g. Den Hartog & Brouwer, 1990, p. 84); sorcerers are evil specialists. All events, especially meteorological phenomena and seasonal changes that escape human control, are under the responsibility of the main god, *Alaona* (Massa), or *Lona* (Mussey). The same name designates the rain, of which he is specifically in charge. He is close to humans during the rains, then goes back to the sky at the onset of the dry season. Lesser deities, who are under his remote control, play different roles. The sun is an evil deity responsible for heat, fever and skin diseases such as leprosy. Mother Earth, 'the earth on which I was born', maintains equilibrium in the village, especially of population size and structure, and of food production. She is therefore closely associated with season. Dead ancestors and a cohort of local clanic guardian spirits are responsible, at a lower level, for events including those due to seasonal variation.

### Religious stresses

There are seasons and specific times of year when issues that are crucial to the community depend on the goodwill of supernatural beings and when their help is awaited with anxiety. Such situations may create mental stress in individuals that requires religious and/or ritual coping activities for its management (e.g. Laughlin & Brady, 1978, p. 29). There are three main religious occasions, related to season: 1. asking for the rains and obtaining good crops; 2. celebrating the community New Year festival; and 3. initiation rites.

Asking for the rains and seeking good crops involves the following observances. When the time comes to sow sorghum, animals are slaughtered, and offerings and prayers are made for the rains. As time passes, the situation grows more tense and the level shifts from praying for rain to fighting drought by placating the deities and counteracting witchcraft. Soul stealing (*fona*) in order to obtain extra shadows (manpower) to work in one's fields is especially frequent at this time. The growth of cereal crops depends on the goodwill of benevolent but still dangerous supernatural beings: God, the ancestors, the village protective spirits and especially Mother Earth. At times, villagers are forbidden from going to the fields (*helda*) by the traditional priest in order to let Mother Earth and other supernatural beings rest, and to allow them to accomplish the difficult task of plant growth and grain maturation. No noise or open conflict should disturb this quiet and subtle process. Among a neighbouring group, the Koma, adultery is severely forbidden during this period. When crops mature, the first fruits of finger millet (*Eleusine*), rain sorghum (*Sorghum caudatum*), and cucumber (*Curcumis sativa*), are collected surreptitiously by the traditional priest in the fields and hung on his doorpost. Thereafter, the tension is relieved. Among the Koma, the time for sexual licence returns as soon as cereals are threshed, beginning with a ritual exchange of obscene songs between the sexes. This contrast in sexual behaviour between the two main seasons is also observed in other tropical societies (e.g. Scaglion, 1978; Condon, 1982, p. 163).

The New Year ritual involves the following practices. Each large Mussey and Massa clan celebrates its specific clan festival, which corresponds to the appearance of the first moon marking the New Year. Each clan celebrates this ritual, *vun tilla* ('the beginning of the months'), according to when it first settled on the land. The clan protectors, Mother Earth, guardian spirits and ancestors, remain near-by in the village. They can be heard and sometimes seen, and they should not be offended by noise, violence or theft. Waiting for the moon to rise is a time of great drama. The appearance of the moon, the linking of a new year to the old one and the perpetuation of life, are believed not to be automatically granted. Ancestors have to be implored to let the moon rise and a New Year to begin: 'Dead father, grandfather, ancestors! Let the moon rise, let it go, don't allow us to remain miserable people!' A burst of glee, accompanied by violent songs celebrating past military successes greet its appearance. The next morning,

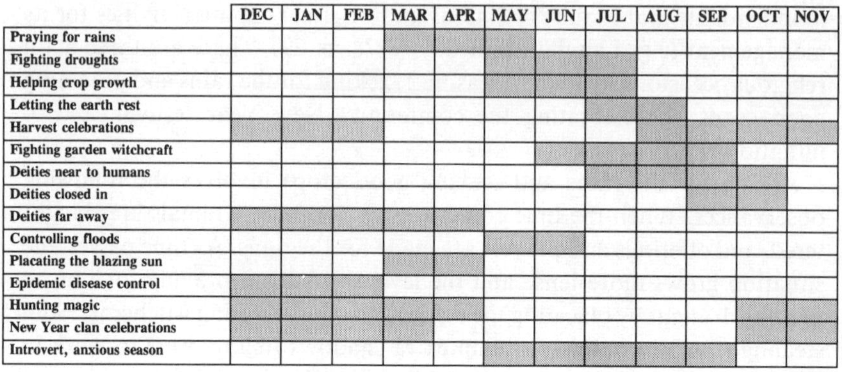

| | DEC | JAN | FEB | MAR | APR | MAY | JUN | JUL | AUG | SEP | OCT | NOV |
|---|---|---|---|---|---|---|---|---|---|---|---|---|
| Praying for rains | | | | ■ | ■ | ■ | | | | | | |
| Fighting droughts | | | | | | ■ | ■ | ■ | | | | |
| Helping crop growth | | | | | | ■ | ■ | ■ | | | | |
| Letting the earth rest | | | | | | | ■ | ■ | | | | |
| Harvest celebrations | ■ | ■ | ■ | | | | | | | ■ | ■ | ■ |
| Fighting garden witchcraft | | | | | | | ■ | | | | | |
| Deities near to humans | | | | | | | | ■ | ■ | ■ | | |
| Deities closed in | | | | | | | | | ■ | ■ | ■ | |
| Deities far away | ■ | ■ | ■ | ■ | | | | | | | | |
| Controlling floods | | | | | | | ■ | ■ | ■ | | | |
| Placating the blazing sun | | | | ■ | ■ | | | | | | | |
| Epidemic disease control | ■ | ■ | ■ | | | | | | | | | |
| Hunting magic | ■ | ■ | ■ | | | | | | | | ■ | ■ |
| New Year clan celebrations | ■ | ■ | | | | | | | | | | |
| Introvert, anxious season | | | | | | | ■ | ■ | ■ | ■ | | |

Fig. 14.3. Seasonal distribution of ritual anxiety.

impurity and respiratory illness (*doyda*) are symbolically chased away to the west, to the nearest neighbours. From then on, the clan celebrates a period of feasting and merriment that lasts between five days (ordinary Mussey clans) and two weeks (the Guisey clan, Massa). Thus a period of intense joy follows an evening of extreme anxiety: the merry dry season has begun.

The initiation ceremony, which sanctions the access of young males to the status of fully-fledged adults, is an ambiguous situation. It is both comforting and status-enhancing, but very stressful psychologically. It is also physically painful when body mutilations such as circumcision are involved. This is not the case among the Massa and Mussey, but it is among the Koma. Fig. 14.3 summarises the times of ritual anxiety across the year.

### Ideals of life and seasonality

To the Massa and Mussey, a 'good life' means being healthy, well fed, happily married and having many offspring. Males should also enjoy prestige and be rich according to local criteria; that is, to be in command of many wives and own numerous cattle. In addition, one should be well protected against supernatural evil and witchcraft.

These are features of lifestyle, long-term situations or personal events, not of seasonality. In general, the biological consequences of affect have been most studied in individuals undergoing permanent or long-term stress due to lifestyle change (James *et al.*, 1989), professional troubles due to a variety of factors, or to shorter-term acute stresses such as mourning (Parkes, 1972), hospital admission, academic examinations, violent films, none of them linked to the relatively short-term effects of seasonality (Harrison, 1982a; James *et al.*, op. cit., p. 284). Indeed, there has been little interest in the study of psychological states linked to seasonality and of their consequences for mental and biological homeostasis, except for 'culture bound' syndromes such as Arctic hysteria (*Pibloktoq*), which occurs primarily in late winter (Wallace, 1972, p. 373; Bohlen, 1979, p. 76).

Verbal expression of emotions by the Massa and Mussey are revealing in relation to seasonal psychological stress. The following traditional concepts are akin to western categories of happiness, sorrow, comfort, stress, and so on: 1. *naa* (Massa), *jifia* (Mussey): 'good' to experience, do or eat. This also means beautiful; 2. *naadi* (Massa), *jifidi* (Mussey), *joo* (Massa, Mussey): 'bad' to experience, do or eat. This also means ugly. The general word for happiness in both languages is *firita*, the opposite of which is *bengyo* (spoilt). A distinction is made between what is encountered by way of misfortune or luck (*halalta* (Massa), *fumsumna* (Mussey)), which can be either good (*cogota*), cooling, placating (*hebe* (Massa), *cede* (Mussey)) or bad, (*joo* (Massa, Mussey)). Another opposition lies between 'leisure' and 'tiring work' – *til luna*, 'the season of play', as against *til zuma* (Massa), *til baratna* (Mussey) 'the season of agricultural toil'.

In the oral tradition, the opening sentence of all tales refers to the spring-like weather at the end of the rainy season: 'Here comes (*falls*) my tale. For me it will be the crop season, I shall drink [fresh] water under the growing grass'. All other themes relate to hunger. Many tales begin with: 'Hunger had fallen heavily on the country. There was nothing at all to eat.' Stealing food and refusing to share, adopting gluttonous behaviour and being punished for it are features of the central character of these tales, *Hlo* (Massa) or *Kada* (Mussey). Proverbs refer to eating, being greedy or selfish. Satirical songs are readily made up to ridicule individuals guilty of food theft, no matter how starved they may have been.

A number of diseases are represented in the local system of divination, among which *doyda* (respiratory diseases, coryza) is associated with season.

Massa and Mussey terms for the calendar months give some objective clues about how they are perceived (Fig. 14.4). In relation to stress, references are made to: 1. hunger, food availability and fatness; 2. cold and dusty mist; 3. disease, colds; 4. heavy field work; and 5. somewhat more originally, to being shut in during the rainy season in order to respect the processes of growth (*helda*) and maturation (*caka*).

### 'Stressful times' as against 'happy times'

The evaluation of seasonal happy and bad times and their contribution toward stress or well-being from behaviour and verbal expression of emotions is not easy, since the appraisal of events and displays of emotion

| | | OCT | NOV | DEC | JAN | FEB | MAR | APR | MAY | JUN | JUL | AUG | SEP |
|---|---|---|---|---|---|---|---|---|---|---|---|---|---|
| SUB-SEASONS | Massa | *CAYTA* = HARVEST | | | | *FATTA* = SUN | | *GAVULTA* = WINDS | | *MENDA* = RAINY SEASON | | *WALLA* = FATTENING * | |
| | Mussey | *SIMETNA* = COLD | | | | *ZAMALLA*= HEAT | | | *BARATNA* = AGR. WORK | | *DOLDA* = RAINY SEASON | | |
| CLIMATE | | WINDY AND COLD | | WINDY & COLD DUSTY MIST | | HOT AND DRY | | | DAMP AND COOLER | | RAINY AND COOL | | DAMP AND WARMER |
| MONTHS | Massa | Cutting stems | Harvest | Adobe making | Cutting thatch | Cutting magic onions | Light Colds | Heavy Colds | Hard Work | Eighth month | Hunger, "Shut In" | Fat * Guru | Very Fat * Guru |
| | Mussey | New Year Ritual names | | Begen. of Cold | Dusty Mist | Hunting places names | | Fishing places names | | "Surr-ounded" | "Closed in" | Where's The Food | Furtive First Fruits |

Fig. 14.4. Massa and Mussey terms for seasons and months referring to stress.
* refers to the individual fattening sessions *'Guru walla'*.

are culture-specific (d'Andrade, 1990, p. 73; de Zalduondo, 1989, p. 533). However, the Massa distinguish two types of objective stress indicators. During the rainy season food shortage, they see signs of starvation in their dogs, which shortly anticipate the villagers' fate. In addition, the sound of angry voices and of children constantly crying are signs of serious food shortage (e.g. Miracle, 1961, p. 278). However, despite food shortages, the Massa and Mussey perceive the rainy season to be a cool and pleasant time, in contrast to the stifling dry-season months. The high incidences of malaria and diarrhoea in the rainy season are not perceived as anything unusual, since both are present all the year round. The cold dry season, a time of plenty and leisure, is however considered to be dangerous, since it is cold and the climate is favourable to disease.

In addition to direct seasonality-related hardships such as hunger and disease, indirect hardships or problems may occur or increase in number or severity as a result of seasonality-bound cultural factors. Drought creates hunger which in turn leads to a much resented increase in theft. The beginning of the dry season is a period of plenty in terms of food and money, a leisurely time suitable for visiting people, travelling, courtship and having love affairs; it also results in a large increase in venereal disease incidence. During the cropping period, good conditions exist for being ambushed by enemies, since people often travel distances to tend to their fields, usually passing through areas where the path is secluded from view by high vegetation. Cold weather from November to January, in a society in which appropriate clothing is lacking, leads to fires being lit in highly inflammable thatched huts. Braziers are kept alight all night long, and often people are badly burned, as a consequence of falling asleep around the fire, and falling in. These are mostly children below the age of four years. At the Gobo dispensary in 1975, 1.5% of all consultations (55 cases) were for burns (Koppert, 1981).

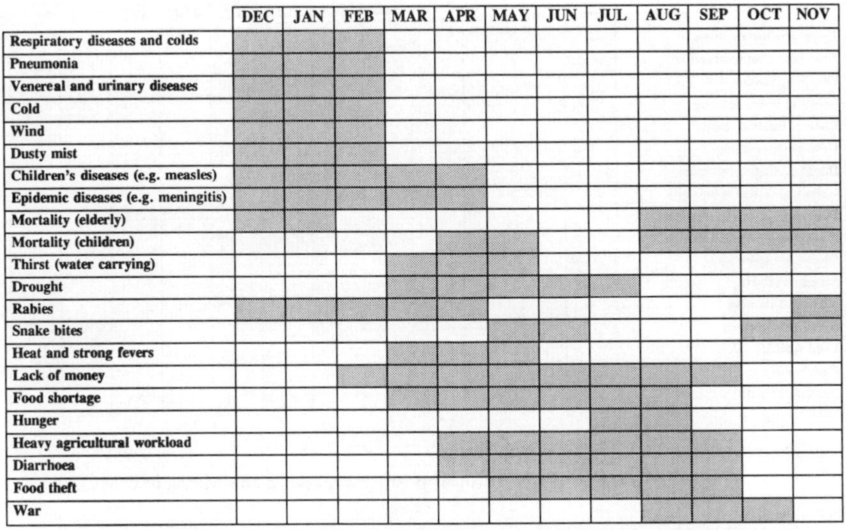

| | DEC | JAN | FEB | MAR | APR | MAY | JUN | JUL | AUG | SEP | OCT | NOV |
|---|---|---|---|---|---|---|---|---|---|---|---|---|
| Respiratory diseases and colds | ▓ | ▓ | | ▓ | | | | | | | | |
| Pneumonia | ▓ | ▓ | | | | | | | | | | |
| Venereal and urinary diseases | ▓ | ▓ | ▓ | ▓ | | | | | | | | |
| Cold | ▓ | ▓ | | | | | | | | | | |
| Wind | ▓ | ▓ | ▓ | | | | | | | | | |
| Dusty mist | ▓ | ▓ | | | | | | | | | | |
| Children's diseases (e.g. measles) | | | ▓ | ▓ | | | | | | | | |
| Epidemic diseases (e.g. meningitis) | ▓ | ▓ | ▓ | ▓ | | | | | | | | |
| Mortality (elderly) | | | | | | | | | ▓ | ▓ | ▓ | ▓ |
| Mortality (children) | | | | ▓ | ▓ | | | | ▓ | ▓ | ▓ | |
| Thirst (water carrying) | | | | | ▓ | ▓ | ▓ | | | | | |
| Drought | | | | | ▓ | ▓ | ▓ | | | | | |
| Rabies | ▓ | ▓ | | | | | | | | | ▓ | ▓ |
| Snake bites | | | | | | | ▓ | ▓ | ▓ | ▓ | ▓ | ▓ |
| Heat and strong fevers | | | | ▓ | ▓ | ▓ | | | | | | |
| Lack of money | | | | ▓ | ▓ | ▓ | ▓ | ▓ | | | | |
| Food shortage | | | | ▓ | ▓ | ▓ | ▓ | ▓ | | | | |
| Hunger | | | | | | ▓ | ▓ | ▓ | | | | |
| Heavy agricultural workload | | | | | ▓ | ▓ | ▓ | ▓ | ▓ | | | |
| Diarrhoea | | | | | ▓ | ▓ | ▓ | ▓ | ▓ | | | |
| Food theft | ▓ | | | | | ▓ | ▓ | ▓ | | | | |
| War | | | | | | | | | ▓ | ▓ | ▓ | |

Fig. 14.5. Seasonal distribution of negative events for the Massa and Mussey.

## Times of year when negative events dominate

As elsewhere in tropical Africa (e.g. Richards & Widdowson, 1936, pp. 179, 189) the rainy season is considered in many ways the most difficult, although the Massa and Mussey also see the pleasant aspects of this time. Fig. 14.5 shows the seasonal distribution of negative events for the Massa and Mussey.

In most years, there is a four week period of food shortage that has little important biological impact, and that does not elicit specific coping behaviours. Traditional terminology distinguishes between various levels of hunger (*mayda*) ranging from the pleasant need to have a meal, through to famine. Only the last stages (*baknarda*: 'tightly-tied skin loincloth', and *may happa*: 'white hunger' (because people have lost their self-respect and no longer wash themselves)) correspond to fear, starvation and helplessness. Mild seasonal food shortages are expected by the Massa and Mussey; these have few biological consequences. Such shortages are a part of domestic life and can be overcome without great difficulty. Nevertheless, the history of the Massa and Mussey is so laden with famine stories that nowadays even a mild food shortage elicits anxiety. Severe shortages bordering on famine may occur in the rainy season of the year following crop failure, when resources are insufficient to provide the food energy needed to perform the work required for a new crop. This happened in 1980. Death from starvation still occurs sporadically during the rainy

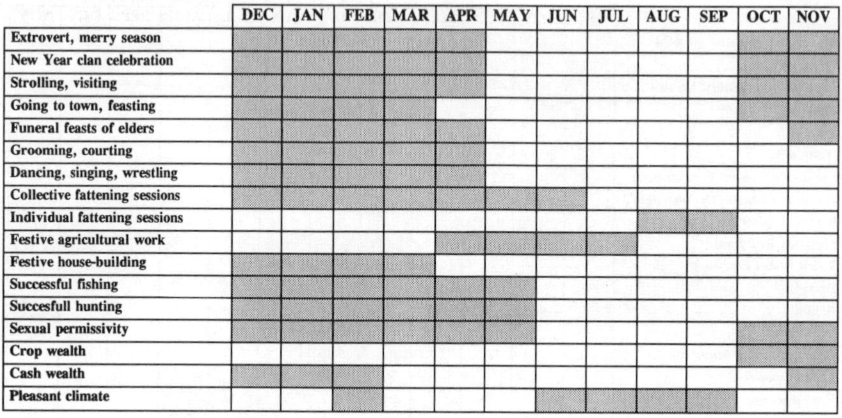

|  | DEC | JAN | FEB | MAR | APR | MAY | JUN | JUL | AUG | SEP | OCT | NOV |
|---|---|---|---|---|---|---|---|---|---|---|---|---|
| Extrovert, merry season | ▓ | ▓ | ▓ |  |  |  |  |  |  |  | ▓ | ▓ |
| New Year clan celebration | ▓ | ▓ |  |  |  |  |  |  |  |  | ▓ | ▓ |
| Strolling, visiting | ▓ | ▓ | ▓ |  |  |  |  |  |  |  | ▓ | ▓ |
| Going to town, feasting | ▓ | ▓ | ▓ | ▓ |  |  |  |  |  |  | ▓ | ▓ |
| Funeral feasts of elders | ▓ | ▓ | ▓ |  |  |  |  |  |  |  |  | ▓ |
| Grooming, courting | ▓ | ▓ | ▓ |  |  |  |  |  |  |  | ▓ | ▓ |
| Dancing, singing, wrestling | ▓ | ▓ | ▓ |  |  |  |  |  |  |  | ▓ | ▓ |
| Collective fattening sessions | ▓ | ▓ | ▓ | ▓ | ▓ | ▓ |  |  |  |  |  |  |
| Individual fattening sessions |  |  |  |  |  |  |  | ▓ | ▓ | ▓ |  |  |
| Festive agricultural work |  |  |  | ▓ | ▓ | ▓ | ▓ | ▓ | ▓ |  |  |  |
| Festive house-building | ▓ | ▓ |  |  |  |  |  |  |  |  |  | ▓ |
| Successful fishing |  |  |  |  |  | ▓ | ▓ |  |  |  |  |  |
| Succesfull hunting | ▓ | ▓ | ▓ |  |  |  |  |  |  |  |  |  |
| Sexual permissivity | ▓ | ▓ | ▓ | ▓ | ▓ | ▓ |  |  |  | ▓ | ▓ | ▓ |
| Crop wealth | ▓ | ▓ |  |  |  |  |  |  |  |  | ▓ | ▓ |
| Cash wealth | ▓ | ▓ |  |  |  |  |  |  |  |  |  |  |
| Pleasant climate |  |  | ▓ |  |  |  | ▓ | ▓ | ▓ | ▓ | ▓ |  |

Fig. 14.6. Seasonal distribution of positive events for the Massa and Mussey.

season, after two consecutive crop failures and when food stores have been low for many months and people have to work hard. This happened after the two crop failures of 1984 and 1985.

Another potentially stressful period is the cold windy and misty period between December and February. Cold, disease and the fear of disease create unrest, although materially and socially the period is a festive one.

The hot season from mid-February to May is also stressful. This is when heat and the supernatural evil influence of the sun are both at their greatest. Agricultural work has begun, and 'working in the sun' is judged to be very exhausting. In the hot, dry environment, biologically stressful temperatures of around 40 °C (Hanna *et al.*, 1989, p. 137) are easily reached, air convection increasing the heat load (Hanna & Brown, 1983, p. 267). Respiratory complaints and sore throats are common, and there is a particular ailment associated with the effects of the sun: *fatta henjeta*, 'sun in the night'. The symptoms include 'a burning body', depigmentation of the skin, and sores on the genitals; these are likely to be sequelae of syphilis and yaws (Buck *et al.*, 1970, p. 191). *Fatta henjeta* generates much emotional unrest, and the help of diviners or witch doctors is needed for its alleviation. Massa and Mussey consider it to be caused by the unescapable, stifling, hot dry season and by working under the blazing sun, being exposed to the evil supernatural powers of the deities of death (*Matna*) and the bush wilderness (*Bagaona*), and to the covert fear of witchcraft.

### Positive affect and well-being

Fig. 14.6 gives seasonal distribution of positive events for the Massa and Mussey. Shweder (1990, pp. 28–9) has proposed six types of relationship between reality constituting psyche (intentional persons) and culturally constituted realities (intentional world), when considering person/environment interactions. Three types of relationship are positive, tending to restore homeostasis. Wood *et al.* (1990, p. 489) suggested that 'high levels of both physical and mental energy are associated with high levels of positive affect . . . High positive affect may lead to feelings of physical competence and mental agility. In addition, a subjective sense of energy is likely to boost both affect and satisfaction with life'. Without entering the debate about the direct biological influence of positive affects (Cabanac, 1986) it is reasonable to suggest that their indirect consequences make individuals dynamic, open to communication with others, and lead to effective action (Folkman & Lazarus, 1988).

The notion of mood is also important in the context of seasonality, being the culturally appropriate state of mind (Cox, 1978; de Zalduondo, 1989, p. 534) to various seasons or seasonal events (Gerber, 1985, p. 130). Among the Massa and Mussey, a mood of restraint dominates during the rainy season; in the dry season, the mood is festive. However, traditional societies may also use joyful moods and mirth to boost morale during difficult periods or when high work performance is required. This is the case among the Massa and Mussey when heavy tasks, mostly field preparation and weeding, have to be rapidly performed (*depma*). On these occasions, special working parties are convened. No money changes hands but the task is performed in a competitive spirit, and abundant food, drink, and eventually girls, are provided at the end. Various forms of festive work have been observed elsewhere in Africa, in South America and Nepal (Panter-Brick, 1989, p. 89 and Chapter 16, this volume).

In a more adventurous way, I have contended that holding the individual *Guru walla* fattening session (de Garine & Koppert, 1991), might be a way in which Massa society demonstrates symbolically that they are in command of seasonal hardship (cf. Grivetti about the Tlokwa, 1981, p. 533). By holding a conspicuous food consumption event at precisely the time when availability is at its lowest, they may be attempting to contradict symbolically what is happening materially in the natural environment. Although food is taken away from the community to feed the *gurna* participants, it might be questioned whether the psychological benefits derived from the successful performance by the community (which could be compared to the village of a winning sports team) compensate for the lack of food or not.

### Conclusion

The anthropological approach has helped to show how seasons and the situations and events linked to them are perceived by two tropical African populations. Detailed ethnographic observation has allowed the times of physical stress to be identified, as well as the times when emotion-laden states, events and behaviour occur. This knowledge may provide the anthropologist with the opportunity to study in a subtle way the possible biological consequences of mental stress aroused by seasonal changes.

### References

Andreoli, A., Taban, C. & Garone, G. (1989). Stress, dépression, immunité: Nouvelles perspectives de recherche dans le domaine de la psycho-immunologie. *Annales de Médécine Psychologique*, **147**(1), 35–46.
Augé, M. (1986). *L'Anthropologie de la Maladie. L'Homme – Anthropologie, Etat des Lieux*. Paris: Navarin, Livre de Poche.
Beauvilain, A. (1989). *Nord Cameroun: Crises et Peuplement*. Tomes I et II. Caen: Alain Beauvilain.
Bénéfice, E. & Chevassus-Agnès, S. (1985). Variations anthropométriques saisonnières des adultes appartenant à deux populations différentes de l'Afrique de l'Ouest. *Revue d'Epidémiologie et Santé Publique*, **33**, 150–60.
Bohlen, J.G. (1979). Biological Rhythms: Human responses in the Polar environment. *Yearbook of Physical Anthropology*, **22**, 47–81.
Boyden, S. (1970). Cultural adaptation to biological maladjustment. In *The Impact of Civilisation on the Biology of Man*, ed. S. Boyden, pp. 190–218. Sydney: Australian University Press.
Brun, T.A., Bleiberg, F. & Gohman, S. (1981). Energy expenditure of male farmers in dry and rainy seasons in Upper Volta. *British Journal of Nutrition*, **45**, 67–82.
Buck, A.A., Anderson, R.I., Sasaki, J.J. & Kawata, K. (1970). *Health and Disease in Chad: Epidemiology, Culture and Environment in Five Villages*. Baltimore: The Johns Hopkins Press.
Butzer, R.W. (1982). *Archaeology in Human Ecology*. Cambridge: Cambridge University Press.
Cabanac, M. (1986). Du confort au bonheur. *Revue de Médecine Psychosomatique*, **6**, 9–15.
Cathebras, P. & Rousset, H. (1988). Esquisse d'une epistémologie du stress. *Revue de Médécine Psychosomatique*, **16**, 61–70.
Chambers, R., Longhurst, R. & Pacey, A. (1981). *Seasonal Dimensions to Rural Poverty*. London: Francis Pinter.
Colby, B.N. (1987). Well-being: A theoretical programme. *American Anthropologist*, **89**, 879–95.
Condon, R.G. (1982). Seasonal variations and interpersonal conflict in the Central Canadian Arctic. *Ethnology*, **XXI**, 151–64.
Cox, T. (1978). *Stress*. London: Macmillan Press.
d'Andrade, R.G. (1984). Cultural meaning systems. In *Culture Theory: Essays on Mind, Self and Emotions*, ed. R.A. Shrewder & R. Alevine, pp. 88–122. Cambridge: Cambridge University Press.
d'Andrade, R.G. (1990). Some propositions about the relations between culture and human cognition. In *Cultural Psychology: Essays on Comparative Human Development*, ed. J.W. Stighler, R.A. Shweder & G. Herdt, pp. 65–129. Cambridge: Cambridge University Press.

de Garine, I. (1980). Approaches to the study of food and prestige in savannah tribes: Massa and Mussey of Northern Cameroon and Chad. *Social Science Information*, **19**, 39–78.

de Garine, I. & Koppert, G.J.A. (1988). Coping with Seasonal Fluctuations in Food Supply among Savanna Populations: The Massa and Mussey of Chad and Cameroon. In *Coping with Uncertainty in Food Supply*, ed. I. de Garine & G.A. Harrison, pp. 210–60. Oxford: Clarendon Press.

de Garine, I. & Koppert, G. (1991). Guru: fattening sessions among the Massa. *Ecology of Food and Nutrition*, **25**, 1–28.

de Zalduondo, B.O. (1989). Ecology and affective behavior: selected results from a quantitative study among Efe foragers of northeast Zaire. *American Journal of Physical Anthropology*, **73**, 533–45.

Etkin, N.L. & Ross, P. (1982). Food as medicine and medicine as food: an adaptive framework for the interpretation of plant utilisation among the Hausa of Northern Nigeria. *Social Science and Medicine*, **16**, 1559–73.

Foeken, D.W.J. (1990). Aspects of seasonality in Africa. In *Research Reports 1990/43*, ed. D.W.J. Foeken & A. den Hartog, pp. 6–29. Leiden, Netherlands: African Studies Centre.

Folkman, S. & Lazarus, R.S. (1988). The relationship between coping and emotion: implications for theory and research. *Social Science and Medicine*, **26**, 309–17.

Fox, R.H. (1953). Energy expenditure of Africans engaged in various rural activities. PhD thesis, University of London.

Froment, A. & Koppert, G. (1989). *La situation nutritionnelle au Cameroun d'après les enquêtes épidémiologiques les plus récentes*. Ministère de l'Enseignement supérieur de l'Informatique et de la Recherche scientifique du Cameroun. Institut de Recherches médicales et d'étude des plantes médicinales, Yaoundé, Cameroun.

Geertz, C. (1973). *Interpretation of Culture*. New York: Basic Books.

Gerber, E. (1985). Rage and obligation: Samoan emotion in conflict. In *Person, Self and Experience*, ed. G. White & J. Kirkpatrick, pp. 121–67. Berkeley: University of California Press.

Grivetti, L.E. (1981). Geographical location, climate and weather and magic: Aspects of agricultural success in the Eastern Kalahari, Botswana. *Social Science Information*, **20**, 509–36.

Gubry, P. (1990). Aspects contemporains de la moralité au Cameroun septen-trional. *Séminaire du groupe Méga-Tchad sur 'La mort dans le Bassin du Lac Tchad'*. Paris: ORSTOM.

Hanna, J.M. & Brown, D.E. (1983). Human heat tolerance: an anthropological perspective. *Annual Review of Anthropology*, **12**, 259–84.

Hanna, J.M., Little, M.A. & Austin, D.M. (1989). Climatic physiology. In *Human Population Biology: A transdisciplinary Science*, ed. M.A. Little & J.D. Haas, pp. 132–51. Oxford: Oxford University Press.

Harrison, G.A. (1982a). Adaptation and well-being. *Proceedings of the Indian Statistical Institute Golden Jubilee International Conference on Human Genetics and Adaptation*, ed. A. Basu & K.C. Malothra, pp. 165–71. Calcutta: Indian Statistical Institute.

Harrison, G.A. (1988). Seasonality and human population biology. In *Coping with Uncertainty in Food Supply*, ed. I. de Garine & G.A. Harrison, pp. 26–31. Oxford: Clarendon Press.

James, G.D., Douglas, E., Crews, D.E. & Pearson, J. (1989). Catecholamines and stress. In *Human Population Biology: A Transdisciplinary Science*, ed. M.A. Little & J. D. Haas, pp. 280–95. Oxford: Oxford University Press.

Kelly, D.H.W. (1966). Measurement for Forearm Blood Flow. *British Journal of Psychiatry*, **112**, 789–98.

Koppert, G. (1981). *Kogoyna, étude alimentaire, anthropométrique et pathologique d'un village massa du Nord Cameroun.* Département de Nutrition, Université des Sciences Agronomiques, Wageningen, Pays Bas. Miméo.

Laughlin, C.D. Jr. & Brady, I.A. (1978). Diaphasis and change in human populations. In *Extinction and Survival in Human Populations*, ed. C.D. Laughlin Jr. & I.A. Brady, pp. 8–48. New York: Columbia University Press.

Lawrence, M., Singh, J., Lawrence, F. & Whitehead, R.G. (1985). The energy cost of common daily activities in African women: the effect of season in dietary studies. *American Journal of Clinical Nutrition*, **42**, 753–63.

Mims, C. (1981). Stress in relation to processes of civilisation. In *The Impact of Civilisation on the Biology of Man*, ed. S. Boyden, pp. 167–89. Sydney: Australian University Press.

Miracle, M.P. (1961). Seasonal hunger: a vague concept and an unexplored problem. *Bulletin de l'IFAN*, **XXIII B**, 271–83.

Pagezy, H. (1986). Seasonal variations in the growth rate of weight in African babies aged 0 to 4 years. *Ecology of Food and Nutrition*, **18**, 29–41.

Pagezy, H. (1988). Contraintes nutritionnelles en milieu forestier équatorial liées à la saisonnalité et à la reproduction: réponses biologiques et stratégies de subsistance chez les Ba-Oto et les Ba-Twa du village de Nzalakenga (Lac Tumba), Zaire. Thèse pour le Doctorat d'Etat ès Science, Université d'Aix/ Marseille III (multi).

Panter-Brick, C. (1989). Subsistence work and motherhood in Salmé, Nepal. PhD Thesis, Oxford University. University Microfilm Int. Ann Arbor, Michigan.

Parkes, C.M. (1972). *Bereavement: Studies in Grief in Adult Life.* New York.

Pasquet, P., Froment, A., Brigant, L., Koppert, G.V.A., Bard, D., de Garine, I. & Apfelbaum, M. (in press). Massive overfeeding in Man (Guru walla): energy cost of body weight gain. *American Journal of Clinical Nutrition.*

Random House (1979). *Dictionary of the English Language.* Random House: New York.

Richards, A.I. & Widdowson, E.M. (1936). A dietary study in Northeastern Rhodesia. *Africa*, **IX**(2), 166–96.

Rosetta, L. (1988). Seasonal variations in food consumption by Serere families in Senegal. *Ecology of Food and Nutrition*, **20**, 275–86.

Scaglion, R. (1978). Seasonal births in a Western Abelam village, Papua New Guinea. *Human Biology*, **50**, 313–23.

Schleifer, S.J., Keller, S.E., Meyerson, A.T., Raskin, M.J., Davis, K.L. & Stein, M. (1984). Suppression of lymphocyte stimulation following bereavement. *Journal of the American Medical Association*, **250**, 374–7.

Schultink, J.W. (1991). *Seasonal Changes in Energy Balance of Rural Beninese Women.* Netherlands: Agricultural University of Wageningen.

Selye, H. (1956). *The Stress of Life.* London: Longmans.

Shweder, R.A. (1990). Cultural psychology, What is it? In *Essays on Comparative Human Development*, ed. J.W. Stigler, R.A. Shweder & G. Herdt, pp. 1–46. Cambridge: Cambridge University Press.

Simons, R.C. & Hughes, C.C. (1985). *The Culture Bound Syndrome*. New York: D. Reidel.

Stigler, J.W., Shweder, R.A. & Herdt, G. (1990). *Cultural Psychology. Essays on Comparative Human Development*. Cambridge: Cambridge University Press.

Teokul, W., Payne, P. & Dugdale, A. (1986). Seasonal variations in nutritional status in rural areas of developing countries: a review of the literature. *Food and Nutrition Bulletin*, **8**, 7–10.

Wallace, A.F.C. (1972). Mental illness, biology and culture. In *Psychological Anthropology*, ed. F.L.K. Hsu, pp. 363–402. Cambridge, MA: Schenkman Publishing.

Widdowson, E.M. (1951). Mental contentment and physical growth. *Lancet*, **i**, 1316–18.

Wolf, S. (1965). *The Stomach*. Oxford University Press: Oxford.

Wood, C., Magnello, M.E. & Jewell, T. (1990). Measuring Vitality. *Journal of the Royal Society of Medicine*, **83**, 486–9.

# 15 *Agriculture, modernisation and seasonality*

REBECCA A. HUSS-ASHMORE

## Introduction

Seasonal fluctuation is an obvious and important aspect of most human environments, producing changes not only in climate, but also in patterns of food intake, birth and conception, child growth, and disease (Chambers *et al.*, 1981). Despite increasing research, the linkages between environmental seasonality and the seasonality of human responses are not well understood. Most societies have devised a range of counterseasonal strategies, including patterns of storage, exchange, and labour allocation (Colson, 1979; Messer, 1989). The scheduling of these economic activities affects in turn the timing and intensity of resource availability and exposure to environmental stressors. Potential biological outcomes include changes in health, nutrient reserves, and work capacity of both individuals and populations.

While human biologists have emphasized the adaptive nature of responses to climatic fluctuation, it is also important to understand the conditions under which seasonal strategies fail. History and tradition, as well as poverty and lack of power, may limit the repertoire of responses that any group can mobilize to deal with seasonality. Further, social and economic change may render traditionally adaptive strategies ineffective. This chapter addresses the impact of economic change in developing countries on the ability of traditional farmers to cope with seasonality. Because much of the impact of seasonality in developing countries is tied to the process of food production, the emphasis here is on seasonal and nutritional changes induced by agricultural development.

Traditional farmers in seasonal environments have a broad spectrum of strategies for coping with environmental and resource fluctuation. Agricultural development has paid little attention to these risk-aversive strategies, concentrating instead on technological change and increased production. Although modernization has the potential to alter workload, diet, and pathogen exposure, there is presently little research assessing the seasonality of health and nutrition in the wake of agricultural development.

Because new methods of agricultural production may interfere with traditional seasonal responses, this chapter first outlines a series of traditional agricultural strategies and the aspects of agriculture that development seeks to change. Evidence for seasonal impacts and nutritional outcomes of agricultural development is then reviewed. Finally, a conceptual framework is presented for integrating changing economic seasonality with human biological outcomes.

## Traditional agricultural strategies

A variety of authors have noted the similarity of strategies among traditional rural societies for coping with scarcity, whether annually recurring or more irregular (Campbell, 1990; Colson, 1979; Curry, 1989; Fleuret, 1986; Messer, 1989; Watts, 1986, 1988). Many of these strategies are designed to even out fluctuation, to distribute it across larger numbers of people, or to avoid the worst effects by moving to another area. Some are employed year-round; others are brought into play only when resources decline. One of the most common and effective techniques for preventing disaster is the maintenance of diversity in food procurement and income strategies.

Traditional agriculture is one of a range of strategies for management of environmental resources. Management techniques within seasonal habitats vary in intensity from the foraging of wild plants as they become available to hydroponic and greenhouse gardens that produce year-round. Each of these produces a characteristic pattern of intra-annual variation in labour inputs, as well as kinds and amounts of foods available. Traditional subsistence systems usually employ a variety of management techniques of differing degrees of intensity. Thus, for example, seasonal rainfed agriculture may be combined with periodic hunting or fishing, or with dry-season cultivation of irrigated or swamp-grown vegetables. Animal management strategies are similarly varied, from harvesting of wild species to stall-feeding of improved livestock breeds.

Counterseasonal strategies employed by traditional farmers include production strategies, storage strategies, consumption strategies, and a range of supporting or auxiliary mechanisms. All of these are subject to change, either intentionally or unintentionally, as a result of development activities.

### Production strategies

Diversified production methods include intercropping, staggered planting, irrigation, and the development of specialized cultivars. Varietal selection and development has traditionally been of paramount importance in this area, and is one of the major emphases of modern agricultural research. For plants, growing season is limited by temperature, rainfall, and photoperiod or daylength. Plants differ widely in their requirements with respect to these three variables, both between and within species. Crops with different planting seasons and different maturation times can be used to even out demand for labour and can provide food during periods of shortage prior to the main harvest. Gill (1991) notes that traditional farmers are well aware of the inverse relationship between maturation rate and biomass in crops, but may choose to forego increased yield for early ripening. Intercropping is another way of providing temporally spaced output from a single plot of land and of consolidating labour tasks (Norman, 1974).

Where plant growth is seasonally limited by rainfall, irrigation or water management can lengthen the growing season and substantially increase output. In Africa, both flood-recession cultivation and swamp cultivation are traditional strategies for extending the growing season. Richards (1985) describes how West African farmers evolved their own system of wet-rice cultivation in valley bottoms as a supplement to their main rainfed crop. These early season crops were used to ride out the hungry season and lowered the severe labour peaks otherwise seen. Other traditional water management techniques have included mulching, ridging of fields, and the storage of water in ponds.

Complementarity in the timing of inputs and outputs for different enterprises is a widely used method of evening out seasonal fluctuation (Gill, 1991). This may mean growing successive crops of different cultivars in the same field, such as millet followed by maize or peas, or rice followed by wheat. It may also mean the use of cultivars with different growth habits or life-spans. Combining annual grains with perennial tree crops or root crops can be one method of spreading both labour requirements and the risk of crop failure. Moris (1989) notes the importance of root crops such as cassava, taro, and sweet potatoes in countering seasonality both for Africa and for Latin America. Bananas and other tree crops also make a substantial contribution to tropical diets and often have a growth cycle quite different from annual crops. Leakey (1986) and Chambers & Longhurst (1986) point out the multiple counterseasonal uses of trees. Besides slack-season fruits and nuts, trees provide oil, starch, edible leaves, cooking fuel, and animal fodder.

The combination of cropping with animal husbandry is a widely recognized method of maintaining complementarity in farming systems. This may be achieved in two ways, either by mixed farming of crops and livestock on the same farm or by specialization and exchange, as among grain farmers and pastoralists. The first has the advantage of evening out the utilization of family labour, in that peak labour periods for crops and for livestock fall at different seasons of the year (White, 1986). Both family diet and income can be supplemented by animal products during the rainy season when grains are scarce.

While pastoralism is a more specialized strategy than mixed farming, and thus more risky in seasonal habitats, the mobility of animals has made it a stable seasonal response. Animals in pastoral systems use a succession of habitats, moving from areas of poor forage to better areas as rainfall changes. Moris (1989) points out that herding animals, rather than pasturing, keeps them from concentrating on favourite species and maintains plant diversity. In addition, mixed herds with different life-spans, breeding seasons, and lactation periods provide flexibility to the system. The worst season for pastoralists tends to occur at the end of the dry season, when pasture and water are both in short supply (Swift, 1981).

While many production strategies concentrate on the timing of outputs, traditional cultivators have also had to plan for the seasonality of inputs. For highly seasonal farming systems, peak labour requirements become a limiting factor in production. Many of the methods cited above are aimed at spreading labour requirements more equitably over the year. Farmers have employed a variety of devices for recruiting and maintaining labour during peak seasons; in addition to family labour, exchange, hired, and community labour have all been important ways to increase manpower supply (Stone *et al.*, 1990).

### Storage strategies

One of the chief differences between foraging societies and those of traditional cultivators is in the amount and kind of food storage. Storage properties have undoubtedly been important in choice of traditional cultivars, and a variety of processing techniques such as drying and fermenting have improved 'shelf life' of stored crops. Brenton (1989) suggests that storage options can be categorized as: 1. technological storage (granaries, baskets, pits); 2. biological storage (human energy reserves); 3. social storage (food exchange); and 4. environmental storage (livestock, wild plants and animals). While traditional agricultural societies have made use of all these options, technological storage is the option most likely to be amenable to significant further increases in efficiency.

### Consumption strategies

In many ways, food consumption strategies are the mirror of storage strategies. Payne (1989) discusses a variety of food consumption strategies for seasonal environments, taking into account the effectiveness of technological storage options. Farmers may choose to balance intake and expenditure across the year, they may try to maintain consumption relatively constant, or they may choose to eat more in the post-harvest period, thus converting technological storage to biological storage. Where post-harvest food losses are high, he suggests, post-harvest feasting may be a rational strategy.

Supplemental food consumption strategies include seasonal consumption of less-preferred foods, seasonal hunting or foraging, consumption of edible weeds (a by-product of crop weeding and cultivation), and purchased foods. Gifts of food are often seasonal and may be linked to major festivals, as may ritual seasonal fasting. Finally, intrahousehold food distribution may change as a result of changing workloads and food availability. Food may be preferentially targetted to productive adults (Gross & Underwood, 1971) or to more vulnerable members such as children (Leonard, 1991; Wheeler & Abdullah, 1988) as seasonal supplies change.

### Supplementary strategies

While a large number of other seasonal strategies have been reported for rural societies, the most important of these are probably non-agricultural income generation and demographic regulation of the household. Many farm households engage in income generation in the slack season, becoming traders, craft producers, food processors, or migrant labourers (Curry, 1989). Income derived from these sources may be critical for households whose agricultural production does not meet their needs. In addition, migrant labourers change the demographic structure of their households by removing consumers during a portion of the year. Other methods of altering household size and structure include fosterage of children and extended visiting among relatives in different resource zones.

### Agricultural development strategies

In contrast to traditional farming, agricultural development has concentrated on increasing overall production, primarily through the development of improved agricultural technology. This emphasis on production has meant that issues of storage, distribution, and consumption have received much less attention within agricultural research institutions. In addition, the emphasis on technology has often obscured the impacts of change on labour organization and supply. Green revolution packages have often assumed an abundance of labour, despite relative seasonal shortages (Woodhouse, 1988). While increased production should mean greater food security in the long run, the period of agricultural transition may increase seasonal risk for resource-poor rural households.

### Production changes

Most technical packages introduced to farmers are aimed toward specialized commercial crop production within a particular agro-ecological zone (Moris, 1989). Crops are bred for higher yields and decreased sensitivity to daylength (Herdt, 1989). Such improved varieties often require increased moisture and fertilizer and are less tolerant of competition. Weeding and plant spacing thus become important, making these crops more labour and capital intensive than traditional varieties. While such packages have sometimes increased mean output per unit of land, they have not necessarily increased output per unit of labour.

Seasonal changes due to production technology are most easily demonstrated for Asia. Herdt (1989) indicates that expanded irrigation, double-cropping, and new plant varieties have changed the timing of the rice harvest in both Indonesia and The Philippines. In both cases, the harvest comes earlier now than it did in the 1950s and is more evenly spread across the season. Despite extended growing periods, Herdt (1989) notes the increased use of labour during *both* wet and dry cropping seasons as irrigation has expanded in The Philippines.

In Thailand, expansion of irrigation schemes in the past 20 years has dramatically increased rice yields (Mongkolsmai & Rosegrant, 1989). Growing rice exports and government pricing policies have lowered seasonal fluctuations in grain prices and availability. For irrigated farms, more intensive cultivation methods have increased both agricultural income and labour demands.

In Africa, formal irrigation schemes have concentrated on large-scale infrastructure such as dams and reservoirs and on the production of rice, sugar, and cotton (Moris, 1989). While irrigated sorghum production has increased food security in the Sudan, the FAO (1986) has reported that most such systems in Africa perform poorly and are not cost-effective. In Swaziland, however, seasonality of supply and demand for vegetables has led to irrigated dry-season vegetable schemes (Curry *et al.*, 1991). Unlike many African irrigation projects, these are small-scale cooperative projects, with much of the labour supplied by women. Richter (1987) found that women involved in irrigated vegetable production increased their incomes and decision-making powers, but spent more time in agriculture, both daily and throughout the year, than non-irrigating counterparts.

For livestock systems, development has often meant the introduction of livestock with higher productivity in terms of meat and milk. Many of these are European breeds with poor tolerance of tropical conditions, requiring high-quality forage and increased veterinary care. Production of fodder crops and year-round pasturing or stall-feeding are also promoted to maintain growth and milk output and to lower disease risk (ILCA, 1987; ILRAD, 1990). Such systems are likely to change the seasonality of time allocation, especially for household members involved in daily livestock care.

In both Africa and Asia, the seasonality of labour supply is a critical variable in the ability of farmers to intensify production. In Asia, such labour demand is seen as potential employment for the landless. In many cases, women and children have been increasingly recruited to fill the gap. Moris (1989) notes that the removal of large numbers of children to full-time schooling has accelerated the shift to hybrid maize production in parts of Africa, since children are not available to scare birds from vulnerable crops such as sorghum. Where development planners have recognized labour constraints, mechanization in the form of tractors, animal traction, or herbicides has been suggested as the solution. Such solutions are often capital-intensive and may simply shift the timing of labour bottlenecks from planting or weeding to the harvest season (Gill, 1991).

### Storage changes

Bates (1986) has noted the relative lack of attention to post-harvest factors in the food chain. Without improvements in food preservation, transport, and marketing, he argues, production increases can have little counter-seasonal effect. Current development efforts concentrate on the construction of pest-resistant containers, particularly for storage on small farms, and more efficient processing for converting crops to storable forms.

Improved marketing and pricing policies have reduced seasonal fluctuations in some cases (Sahn & Delgado, 1989). Even so, many farmers continue to 'sell low' in the post-harvest season and 'buy high' in the pre-harvest period of low supply (Ellsworth & Shapiro, 1989). In Burkina Faso, wealthier households were able to take advantage of seasonal price swings in their marketing of grain, while poorest households were most disadvantaged by the market. In general, goverment policies to improve predictability of food and commodity prices seem to have succeeded better in Asia than in Africa (Herdt, 1989).

### Consumption changes

Agricultural development can alter the seasonality of consumption by changing timing and control of income (farm and non-farm), by changing crop mix and the production of minor crops, or by changing food processing and meal production. Strategies that provide regular income to women are most likely to improve household food consumption (Frankenberger, 1985; Huss-Ashmore, 1992). While development of secondary crops with staggered harvesting time can lower fluctuation in diet, introduced technology may also remove such crops from the farming system. Improved grain varieties are less tolerant of intercrops, and mechanized production may require the removal of tree crops from fields and borders. In Swaziland, the Cropping Systems Project found that increased use of herbicides on maize would threaten the supply of edible weeds, a staple of the pre-harvest diet (Huss-Ashmore & Curry, 1989a). As a result, project agronomists recommended that herbicide application be delayed until peak palatability of weeds had passed.

Increased processing requirements are a potential hazard of introduced food varieties, in that grain varieties bred for mechanical harvesting may require more time to thresh, grind, and cook. Increased labour requirements of intensified cropping systems also reduce time available to women for fuel collection and cooking (Jiggins, 1986). On the other hand, women's increased involvement in agricultural development may also lead to changing food allocation within the household. For example, Behrman & Deolalikar (1989) suggest that parental strategies during the lean season favour productivity over equity, giving productive members a nutritional advantage.

Finally, food aid and nutritional interventions have been suggested as alternatives and accompaniments to agricultural development. While targeted interventions can improve biological status seasonally (Lawrence *et al.*, 1984, 1989), general food aid is increasingly seen as a disincentive to increased production and long-term food security (Robins, 1991).

### Changes in other strategies

Despite its income-generating potential, commercial agriculture has not always reduced dependence on wage labour. Gill (1991) and Mongkolsmai & Rosegrant (1989) show that labour migration may increase for some portions of the population with intensified cropping schemes. Seasonal timing of wage labour may change, however. As Curry (1989) argues for the Hausa of Niger, an increasing commercial orientation has resulted in extending migration beyond the traditional dry-season period. Low (1986) further argues that improved agricultural technology may actually be adopted in order to free household labour for wage employment. Using Swazi data, he shows that households use modern inputs to increase maize *productivity* (kg/ha), but not total production. After subsistence goals are met, both male and female labour saved is used for off-farm employment.

**Nutrition and agricultural development**

The impact of agricultural development projects has generally been measured in terms of increased crop yields or increased income. Recent recognition that income is unequally shared and imperfectly translated into household welfare has led to suggestions that nutrition and food consumption might be used as outcome variables (Frankenberger, 1985; Martorell, 1982; Mokbel & Pellett, 1987). While a number of authors have suggested that agricultural development is likely to have negative impacts on food consumption (Fleuret & Fleuret, 1980), the available evidence is mixed. This is not surprising, in that few studies have measured impact in the same way, and even fewer have examined change in the same group of farmers over time.

Von Braun & Kennedy (1986) reviewed a series of studies on cash-cropping to determine the effects on either food consumption or nutritional status. Of 14 studies, seven reported positive impacts, five reported negative impacts and two reported no significant change. There was little agreement on the effects of technological change or of crop types, although sugar cane was most likely to have negative outcomes. While intensive production of non-food cash crops is thought to involve more nutritional risk than production of food crops, subsequent studies suggest that nutritional effects of commercial farming may be time and place specific.

Household nutritional outcomes are a function of *total* economic strategies, including both crop production and other income sources. Kaiser & Dewey (1991), for example, showed that the proportion of income from cash cropping had no relationship to child nutritional status for rural households in Guanajuato, Mexico. However, reliance on sporadic migrant wages was negatively associated with nutritional status, as was seasonal illness. In addition, Immink & Alarcon (1991) suggest that different work requirements may explain the varying impacts of commercial cropping for small farm households in western Guatemala. Households growing labour-intensive potatoes as a sale crop were most likely to suffer adverse income and nutritional effects.

Pinstrup-Andersen & Jaramillo (1989) have investigated the impact of technological change in rice production on food consumption in North Arcot district, India. Between 1973 and 1983, household energy consumption increased by 65% and protein consumption doubled, with small farmers showing higher proportional increases than large farmers. However, seasonal fluctuation in both income and consumption also increased over the study period. In this case, improved production technology appears to have raised consumption above minimum levels, even at the lean season, without removing seasonality.

Research in Africa indicates that diversified strategies, including both commercial agriculture and wage employment, are important for food security at the household level. For commercialized farmers in Rwanda, both higher income and greater subsistence production are associated with reduced stunting in children (Von Braun *et al.*, 1991). Poor nutritional outcomes in resource-poor households are linked to time constraints on women that prevent their wage employment and pull children into domestic labour.

Wage employment has also been shown to be a critical variable in the nutritional success of commercializing farmers in Kenya. Studies of nutrition in Coast Province, Kenya, indicate that children on settlement schemes have better diets and show better growth than children in the general rural population, and that children of commercial dairy farmers do even better (Niemeijer *et al.*, 1988a). The best nutritional status, however, is seen among rural children whose families have a full-time wage worker and whose household income is steady (Leegwater *et al.*, 1990). The impact of dairy development is significant for rural coastal households, in that demand for dairy products locally and in the tourist industry is high. Using data from the nutrition studies cited here, Huss-Ashmore (1992) showed that one improved dairy cow could provide sufficient annual income to purchase approximately 300 person-days of a balanced diet at local prices.

In contrast to Asia, irrigated rice production in Africa may not have a positive effect on food consumption. Niemeijer *et al.* (1988b) examined food intake and nutritional status for tenant farmers resident on a Kenyan rice-production scheme, comparing them with non-resident tenants with land outside the scheme and with independent producers. Non-resident tenants, while they sold their rice-crop to the irrigation scheme, had the broadest resource spectrum and best nutritional outcomes. Worst nutritional status was seen among least diversified producers, i.e., rice producers resident on irrigation project land.

In southwest Kenya, nutritional status of small-holders producing sugarcane commercially has been compared with non-sugar producers and with wage-earners and merchants in the local area (Kennedy & Cogill, 1987). Increased sugar income was largely spent on non-food items, and although energy intake was higher, nutritional status of preschool children in sugar-growing households was not better than in other groups studied. Children's nutritional status was related more to morbidity than to income, with 50–70% of women and children in the study ill at any time. Children in female-headed households did better than others, probably due to greater utilization of health care for sick children (Kennedy & Cogill, 1987).

For this study, food intake, nutritional status, and morbidity were measured at four different times during the year. Household intake during

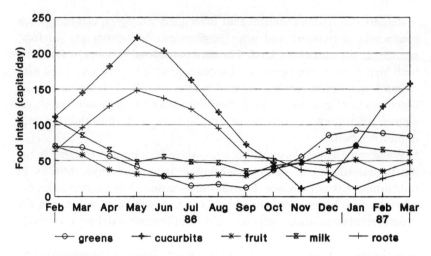

Fig. 15.1. Monthly per capita intake of selected foods by Swazi women in households associated with the Swaziland Cropping Systems Project, 1986–87.

the harvest and post-harvest seasons was significantly higher than during the pre-harvest season. For both women and children, weight-for-height was lowest during the pre-harvest season. Total illness, but not diarrhoeal illness, was seasonal in pattern, with increases during the pre-harvest season. While sugar households had less diarrhoea than other farmers, all farm households suffered more illness than non-farming households in the community (Kennedy & Cogill, 1987).

Finally, research in Swaziland has shown the nutritional impacts of different production orientations (Huss-Ashmore & Curry, 1989b). Farmers emphasizing the production of hybrid maize exhibited higher food intakes and higher body weights than farmers concentrating on cotton production. Stunting (low height-for-age) was less frequent among pre-school children in maize-producing households, but was also related to rainfall, a proxy for the disease environment. Maize-growing households also had more adults in the household, a higher producer–consumer ratio, and more adults (including women) employed off-farm. Energy and protein content of diets was lowest in the pre-harvest season (October–January) and highest during the harvest (February–May). Dietary variety was generally low during the spring (Fig. 15.1), when wild greens, pumpkins, root crops, fruit, and milk were all in short supply.

### Conceptual framework

Taken together, these studies suggest that agricultural development may increase incomes and food consumption, but does not do away with their seasonality. The goals of agricultural development are seldom phrased in seasonal terms, and thus have less to do with countering seasonality than with increasing overall output. Where seasonality has been recognized, it is assumed that raising mean income levels will prevent consumption from falling below some critical threshold, even in lean seasons.

Rather than reducing peak labour demand, agricultural intensification appears to increase overall labour requirements, especially in the case of irrigation. As various authors point out (e.g. Gill, 1991), marginal returns to labour may be evened out across seasons, primarily by lengthening growing period and by raising labour demand during the dry season. This demand may be filled by hiring landless labourers, where these exist, or by the recruitment of women and children to commercial cropping.

Where markets operate efficiently and storage is well-developed, seasonal swings in food prices and access may be reduced. The need for cash continues to be seasonal, however, due to changing needs for agricultural inputs, school fees, and other social obligations across the year. Farmers thus continue to diversify their options by recourse to non-farm income, including wage labour.

While some of the economic outcomes of agricultural development can be predicted, the biological outcomes are far less clear. Growth and nutritional status have rarely been studied in the context of agricultural development and seasonal linkages remain essentially unexplored. The studies reviewed here suggest that disease may be as important to overall biological status as seasonal changes in food supply. We know that agricultural development can increase exposure to disease, including schistosomiasis, trypanosomiasis, and onchocerciasis, (e.g., Ebene, 1987; Hecklau, 1987; McMillan, 1989) as well as brucellosis and tuberculosis. Increased income can increase access to medical care, but time spent on caring for the sick also acquires a greater opportunity cost, especially at peak labour periods. Loss of labour due to illness and care of the sick may limit a family's ability both to intensify production and to replace lost production with wages.

Table 15.1 presents a framework for exploring the impact of agricultural development on biological outcomes. This framework uses a 'proximate determinants' or 'intermediate variables' approach, analogous to those developed within demography and public health. Such approaches have proven to be powerful conceptual tools for guiding research on factors influencing both fertility and child survival. Within this framework, social and economic factors can affect biological outcomes only through their impact on a specified number of intervening variables.

Agricultural projects frequently aim to alter the process of commodity production (Table 15.1). While improved production technology is often

Table 15.1. *Variables linking agricultural development to seasonality and biological outcomes*

| Agricultural development changes: | Subsequent seasonal changes include: | Seasonal changes operate through: | Biological outcomes include effects on: |
|---|---|---|---|
| PRODUCTION<br>Technology<br>Labour organization<br>Land use<br>Credit | TIME ALLOCATION<br>Social maintenance<br>Work (domestic,<br>wage, agriculture) | NUTRITION<br>Food intake<br>Nutrient<br>requirements | NUTRIENT RESERVES<br><br>GROWTH<br><br>WORK CAPACITY |
| STORAGE AND<br>MARKETS | FOOD PRODUCED<br><br>INCOME | HEALTH<br>Pathogen exposure<br>Injury | MORBIDITY<br><br>MORTALITY |
| INFRASTRUCTURE<br>Transport<br>Water storage<br>and management | EXPENDITURE<br><br>ENVIRONMENTAL<br>CONDITIONS<br>Water supply | Health care | REPRODUCTION AND<br>LACTATION |
| IDEOLOGY | (amount, quality)<br>Pesticide use<br>Vector ecology | | |

the focus of this process, altered land use and productive organization may also be recommended. Increased credit to facilitate the adoption of new technology may also be part of the development package. Storage and marketing have received less attention from agricultural planners, but are important elements in shaping food policy. Finally, agricultural development is concerned with changing farmers' knowledge and ideology, in order to develop a commercial orientation.

These direct goals of agricultural change affect a wide variety of economic and environmental factors. Economic factors include time allocation, income, expenditure, and subsistence production. Changes in these variables may take the form of altered scheduling, as in the case of time use, or altered magnitude or amounts of response (income, expenditures), or both. Thus, in the case of labour involved in cash cropping, development might change both the seasonal pattern of labour demand and the total amount of time and energy required. Environmental conditions may also change, for example, if introduced technology results in increased water supply, improved water quality, increased vector survival, or increased pesticide residues.

The biological outcomes of these economic and environmental changes are mediated through changes in nutrition and health. As Table 15.1 indicates, nutritional variables include food intake and nutrient requirements, both of which vary with seasonal changes in workload and food supply. Pathogen exposure, injury, and the availability of care will also vary with changing agricultural operations. These variables can be seen as proximate determinants of biological impact. Impact can then be measured in terms of nutrient reserves, growth, work capacity, morbidity, and mortality. For women, reproductive outcomes and lactation performance would also be influenced by nutrition and health variables.

The use of a conceptual framework such as this could improve our understanding of both the short-term and long-term impacts of agricultural change. In the past, dramatic changes in subsistence systems appear to have been accompanied by increased disease and mortality risk to human populations. Archaeologists and palaeopathologists suggest that negative outcomes were especially likely during the initial period of economic transition (e.g., from foraging to agriculture) (Cohen & Armelagos, 1984). The period of transition to commercial cropping seems also to be risky for modern small-scale farmers, as traditional networks and coping strategies are replaced by immature and inefficient state-level institutions (Sahn, 1989). Biological outcomes are sensitive indicators of both the success and the cost of this change.

From a biological point of view, any population has two adaptive tasks, to procure sufficient resources to meet biological and social needs, and to avoid environmental stressors that might cause biological disruption. The most successful population should be the one that does these tasks most effectively and most efficiently, incurring the least cost. In a seasonal environment, this should mean coping with the most stressful season in ways that maintain health and biological function overall. Within this framework, agricultural development can be seen as one of an array of strategies for procuring resources. Its adaptive success can be judged by its ability to provide adequate nutritional and other resources during the lean season, while not jeopardizing health, biological function, or long-term productive capacity.

### References

Bates, R.P. (1986). Postharvest considerations in the food chain. In *Food in Subsaharan Africa*, ed. A. Hansen & D. McMillan, pp. 239–53. Boulder, CO: Lynne Rienner.
Behrman, J. & Deolalikar, A. (1989). Agricultural wages in India: the role of health, nutrition, and seasonality. In *Seasonal Variability in Third World Agriculture*, ed. D. Sahn, pp. 107–17. Baltimore: Johns Hopkins University Press.
Brenton, B. (1989). The seasonality of storage. In *Coping with Seasonal Constraints*, ed. R. Huss-Ashmore with J.J. Curry & R.K. Hitchcock, pp. 45–55. Philadelphia: MASCA, University Museum, University of Pennsylvania.

216    *Rebecca A. Huss-Ashmore*

Campbell, D.J. (1990). Strategies for coping with severe food deficits in rural Africa: a review of the literature. *Food and Foodways*, **4**, 143–62.

Chambers, R. & Longhurst, R. (1986). Trees, seasons and the poor. In *Seasonality and Poverty*, ed. R. Longhurst. *IDS Bulletin*, **17**, 44–50.

Chambers, R., Longhurst, R. & Pacey, A. (1981). *Seasonal Dimensions to Rural Poverty*. London: Frances Pinter.

Cohen, M.N. & Armelagos, G.J. (1984). *Paleopathology at the Origins of Agriculture*. Orlando: Academic Press.

Colson, E. (1979). In good years and bad: food strategies of self-reliant societies. *Journal of Anthropological Research*, **35**, 18–29.

Curry, J.J. (1989). Seasonality, monetization, and occupational diversity in a Hausa village in Niger. In *Coping with Seasonal Constraints*, ed. R. Huss-Ashmore with J.J. Curry & R.K. Hitchcock, pp. 121–30. Philadelphia: MASCA, University Museum, University of Pennsylvania.

Curry, J.J., Huss-Ashmore, R., Grenoble, D. & Gama, D. (1991). Seasonality of vegetable use and production on Swazi Nation Land: problems and interventions. In *African Food Systems in Crisis. Part Two: Contending with Change*, ed. R. Huss-Ashmore & S.H. Katz, pp. 227–44. New York: Gordon and Breach.

Ebene, R.T. (1987). Health risks to the population of Gezira/Sudan following irrigation of the cotton fields. In *Health and Disease in Tropical Africa*, ed. R. Akhtar, pp. 293–304. Chur: Harwood Academic Publishers.

Ellsworth, L. & Shapiro, K. (1989). Seasonality in Burkina Faso Grain Marketing: farmer strategies and government policy. In *Seasonal Variability in third World Agriculture*, ed. D. Sahn, pp. 196–208. Baltimore: Johns Hopkins University Press.

FAO (1986). *Irrigation in Africa South of the Sahara*. FAO Investment Center Technical Paper no. 5. Rome: FAO.

Fleuret, A. (1986). Indigenous responses to drought in sub-Saharan Africa. *Disasters*, **10**, 224–9.

Fleuret, P. & Fleuret, A. (1980). Nutrition, consumption, and agricultural change. *Human Organization*, **39**, 250–60.

Frankenberger, T. (1985). *Adding a Food Consumption Perspective to Farming Systems Research*. Washington DC: Nutrition Economics Group, Technical Assistance Division, OICD/USDA.

Gill, G.J. (1991). *Seasonality and Agriculture in the Developing World*. Cambridge: Cambridge University Press.

Gross, D. & Underwood, B. (1971). Technological change and caloric costs: sisal agriculture in northeastern Brazil. *American Anthropologist*, **73**, 725–40.

Hecklau, H.K. (1987). Regional development and tsetse fly control in the Coast Province of Kenya. In *Health and Disease in Tropical Africa*, ed. R. Akhtar, pp. 335–58. Chur: Harwood Academic Publishers.

Herdt, R.W. (1989). The impact of technology and policy on seasonal household food security in Asia. In *Seasonal Variability in Third World Agriculture*, ed. D. Sahn, pp. 235–45. Baltimore: Johns Hopkins University Press.

Huss-Ashmore, R. (1992). *Nutritional Impacts of Intensified Dairy Production: An Assessment of Coast Province, Kenya*. ILRAD Technical Reports No. 1. Nairobi: International Laboratory for Research on Animal Diseases.

Huss-Ashmore, R. & Curry, J.J. (1989a). Diet, nutrition and agricultural development in Swaziland. 1. Agricultural ecology and nutritional status. *Ecology of Food and Nutrition*, **23**, 189–209.

Huss-Ashmore, R. & Curry, J.J. (1989b). Diet, nutritional status, and economic constraints in rural Swazi households. *American Journal of Physical Anthropology*, **78**, 243.

Immink, M. & Alarcon, J. (1991). Household food security, nutrition and crop diversification among smallholder farmers in the highlands of Guatemala. *Ecology of Food and Nutrition*, **25**, 287–305.

International Laboratory for Research on Animal Diseases (1990). *ILRAD Annual Report*. Nairobi: ILRAD.

International Livestock Center for Africa (1987). *ILCA's Strategy and Long-Term Plan: A Summary*. Addis Ababa: ILCA.

Jiggins, J. (1986). Women and seasonality: coping with crisis and calamity. In *Seasonality and Poverty*, ed. R. Longhurst. *IDS Bulletin*, **17**, 9–18.

Kaiser, L. & Dewey, K. (1991). Household economic strategies and food expenditure patterns in rural Mexico: impact on nutritional status of preschool children. *Ecology of Food and Nutrition*, **25**, 147–68.

Kennedy, E. & Cogill, B. (1987). *Income and Nutritional Effects of the Commercialization of Agriculture in Southwestern Kenya*. Research Report 63. Washington DC: International Food Policy Research Institute.

Lawrence, M., Lamb, W.H., Lawrence, F. & Whitehead, R.G. (1984). Maintenance energy cost of pregnancy in rural Gambian women and influence of dietary status. *Lancet*, **2**, 363–5.

Lawrence, M., Lawrence, F., Cole, T.J., Coward, W.A., Singh, J. & Whitehead, R.G. (1989). Seasonal pattern of activity and its nutritional consequence in the Gambia. In *Seasonal Variability in Third World Agriculture*, ed. D. Sahn, pp. 47–56. Baltimore: Johns Hopkins University Press.

Leakey, C. (1986). Biomass, man and seasonality in the tropics. In *Seasonality and Poverty*, ed. R. Longhurst, *IDS Bulletin*, **17**, 36–43.

Leegwater, P., Ngolo, J. & Hoorweg, J. (1990). *Nutrition and Dairy Development in Kilifi District*. FNSP Report xx (draft). Nairobi and Leiden: Ministry of Planning and National Development, African Studies Center.

Leonard, W.R. (1991). Household-level strategies for protecting children from seasonal food scarcity. *Social Science and Medicine*, **33**, 1127–33.

Low, A. (1986). *Agricultural Development in Southern Africa: Farm-Household Economics and the Food Crisis*. London: James Currey.

Martorell, R. (1982). *Nutrition and Health Status Indicators*. Working Paper No. 13, Living Standards Measurement Study. Washingdon, DC: The World Bank.

McMillan, D. (1989). Seasonality, planned settlements, and river blindness control in Burkina Faso. In *Coping with Seasonal Constraints*, ed. R. Huss-Ashmore with J.J. Curry & R.K. Hitchcock, pp. 96–120. Philadelphia: MASCA, University Museum, University of Pennsylvania.

Messer, E. (1989). Seasonal hunger and coping strategies: an anthropological discussion. In *Coping with Seasonal Constraints*, ed. R. Huss-Ashmore with J.J. Curry & R.K. Hitchcock, pp. 131–41. Philadelphia: MASCA, University Museum, University of Pennsylvania.

Mokbel, M. & Pellett, P. (1987). Nutrition in agricultural development in Aleppo Province, Syria: 1. Farm resources, rainfall and nutritional status. *Ecology of*

*Food and Nutrition*, **20**, 1–14.

Mongkolsmai, D. & Rosegrant, M.W. (1989). The effect of irrigation on seasonal rice prices, farm income, and labor demand in Thailand. In *Seasonal Variability in Third World Agriculture*, ed. D. Sahn, pp. 246–63. Baltimore: Johns Hopkins University Press.

Moris, J. (1989). Indigenous versus introduced solutions to food stress in Africa. In *Seasonal Variability in Third World Agriculture*, ed. D. Sahn, pp. 209–34. Baltimore: Johns Hopkins University Press.

Niemeijer, R., Foeken, D., Okello, W. & Hoorweg, J. (1988a). *Nutritional Conditions at a Settlement Scheme in Coast Province*. FNSP Report xx (draft). Ministry of Planning and National Development. Nairobi and Leiden: African Studies Center.

Niemeijer, R., Geuns, M., Kliest, T., Ogonda, V. & Hoorweg, J. (1988b). Nutrition in agricultural development: the case of irrigated rice cultivation in West Kenya. *Ecology of Food and Nutrition*, **22**, 65–81.

Norman, D.W. (1974). Rationalizing mixed cropping under indigenous conditions: the example of Northern Nigeria. *Journal of Development Studies*, **11**, 3–21.

Payne, P. (1989). Public health and functional consequences of seasonal hunger and malnutrition. In *Seasonal Variability in Third World Agriculture*, ed. D. Sahn, pp. 19–46. Baltimore: Johns Hopkins University Press.

Pinstrup-Andersen, P. & Jaramillo, M. (1989). The impact of drought and technological change in rice production on intrayear fluctuations in food consumption: The case of North Arcot, India. In *Seasonal Variability in Third World Agriculture*, ed. D. Sahn, pp. 264–84. Baltimore: Johns Hopkins University Press.

Richards, P. (1985). *Indigenous Agricultural Revolution: Ecology and Food Production in West Africa*. London: Hutchinson.

Richter, C. (1987). *An Examination of the Effects of an Irrigation Project on the Lives of Rural Women Farmers in Swaziland*. End of Tour Report, Swaziland Cropping Systems Project. Malkerns, Swaziland: Malkerns Research Station.

Robins, E. (1991). The lesson of Rwanda's agricultural crisis: increase productivity, not food aid. In *African Food Systems in Crisis. Part Two: Contending with Change*, ed. R. Huss-Ashmore & S.H. Katz, pp. 245–68. New York: Gordon and Breach.

Sahn, D. (1989). Policy recommendations for improving food security. In *Seasonal Variability in Third World Agriculture*, ed. D. Sahn, pp. 301–16. Baltimore: Johns Hopkins University Press.

Sahn, D. & Delgado, C. (1989). The nature and implications for market interventions of seasonal price variability. In *Seasonal Variability in Third World Agriculture*, ed. D. Sahn, pp. 179–95. Baltimore: Johns Hopkins University Press.

Stone, G.D., Netting, R.M. & Stone, M.P. (1990). Seasonality, labor scheduling, and agricultural intensification in the Nigerian Savanna. *American Anthropologist*, **92**, 7–23.

Swift, J. (1981). Labour and subsistence in a pastoral economy. In *Seasonal Dimensions to Rural Poverty*, ed. R. Chambers, R. Longhurst & A. Pacey, pp. 80–7. London: Frances Pinter.

Von Braun, J. & Kennedy, E. (1986). *Commercialization of Subsistence Agriculture: Income and Nutritional Effects in Developing Countries*. Washington, DC:

International Food Policy Research Institute.

Von Braun, J., DeHaen, H. & Blanken, J. (1991). *Commercialization of agriculture under population pressure: effects on production, consumption, and nutrition in Rwanda.* Research Report 85. Washington, DC: International Food Policy Research Institute.

Watts, M. (1986). Drought, environment, and food security: some reflections on peasants, pastoralists, and commoditization in dryland West Africa. In *Drought and Hunger in Africa*, ed. M.H. Glantz, pp. 171–212. Cambridge: Cambridge University Press.

Watts, M. (1988). Coping with the market: uncertainty and food security among Hausa peasants. In *Coping with Uncertainty in Food Supply*, ed. I. de Garine & G.A. Harrison, pp. 260–89. Oxford: Clarendon Press.

Wheeler, E.F. & Abdullah, M. (1988). Food allocation within the family: response to fluctuating food supply and food needs. In *Coping with Uncertainty in Food Supply*, ed. I. de Garine & G.A. Harrison, pp. 437–51. Oxford: Clarendon Press.

White, C. (1986). Food shortage and seasonality in WoDaaBe communities in Niger. In *Seasonality and Poverty*, ed. R. Longhurst. *IDS Bulletin*, **17**, 19–26.

Woodhouse, P. (1988). *The Green Revolution and Food Security in Africa: Issues in Research and Technology Development.* DPP Working Paper No. 10. Milton Keynes: The Open University.

# 16 *Seasonal organisation of work patterns*

C. PANTER-BRICK

## Introduction

Thirty-five years ago, Charles Erasmus focused attention on the changes occurring in household labour organisation in South America, where wage labour was rapidly supplanting two traditional forms of inter-household labour organisation, namely festive and exchange labour. He saw that rural households relied upon festive labour groups to complete 'unpredictable urgent tasks' ('the result of a delay in farm work caused by some unexpected circumstance, such as illness in the family, irregular rains, occupancy of a farm late in the season, and enforced absence from the farm'; Erasmus, 1965, p. 182). They also relied upon reciprocal labour exchanges where 'predictable urgent tasks conform to seasonal peak labor loads' (for example, 'clearing before the rains come or before the weeds can grow back, weeding before the crop is choked, and harvesting or processing before crop spoilage can occur'; Erasmus, op. cit.). The operative words, which define both the nature of agricultural activities and seasonal constraints, are 'urgent' and 'predictable'.

Extensive literature on the organisation of household labour now exists, but the purpose here is restricted to demonstrating how labour organisation interacts with seasonality. One must go further than describing a calendar of agricultural activities and 'recapitulate the myriad of germane papers that establish seasonality in work requirements exists in all agricultural systems' (Alderman & Sahn, 1989, p. 82). Instead, it will be useful to review the relative merits and drawbacks of different types of labour organisation, in order to evaluate household strategies as responses to a range of socio-economic and geographic environments. In changing environments where a choice of work patterns is possible, one could expect rural households to shift labour strategies with the cyclical change of seasons. Such a shift would demonstrate the existence of a fine tuning between household labour organisation and ecological conditions.

In this chapter, the ways in which communities living in seasonal environments organise themselves to accomplish their subsistence tasks, and the underlying logic governing their choice of work patterns will be examined. Following this discussion, detailed time-allocation data illustrating the relationship between labour organisation and seasonal constraints in rural Nepal will be analysed.

### Schedule of people, time and place

Some of the best studies on labour organisation were done in South America by authors who vividly portrayed the element of urgency in completing given subsistence tasks, as a result of climatic constraints requiring periodic increases in work-load. Brown, documenting exchange labour in two Aymaran communities in the Peruvian Andes, states the problem convincingly:

> we can realistically estimate that the average Aymara household, relying solely upon its own labor (the equivalent of 2.37 full-time adult workers), will need to spend over 160 days in their fields . . . .. The major problem is not simply that the household has spent 160 days in their fields. Rather, the problem is scheduling the activities. During planting season alone, the household – without outside labor inputs – would have to spend 80 days preparing the fields, plowing, applying fertilizer, and planting. It could not possibly complete the work in time for the crop to mature before the onset of freezing temperatures (Brown, 1987, p. 230).

Aymaran households solve this problem by recruiting extra-familial labour at critical times of the year, through reciprocal exchange with their kindred, wage labour, or exchange of a day's work for a portion of the crop harvested. Recruiting manpower outside the household is the logical solution for meeting one's urgent labour requirements.

An obvious question to ask is how households ever manage to recruit other workers if agricultural tasks are quite so urgent for all farmers. As Erasmus first noted:

> the answer is simple – the seasons allow enough leeway that operations on neighboring farms are by no means simultaneous. In the tropics the flexibility of work schedules may be very great . . . the dates of planting may be deliberately staggered by those exchanging labor, and (farmers) even diversified their crops to avoid conflicts in working schedules. In the highlands of western South America the climatic changes may be great enough within short distances that neighbors frequently have slightly different planting and harvesting seasons . . . .. Where irrigation is practiced . . . the rotation of irrigation periods may result in different individual cropping schedules (Erasmus, 1965, p. 182–3).

In other words, inter-household labour exchanges are successful where farmers stagger their activities in time and place, such that all are not 'feeling the labor pinch at the same time' (Brown, 1987, p. 230). Such a system may capitalise on the fact that land and labour resources are never equally distributed, even in so-called egalitarian subsistence communities where there is no pool of landless workers available for recruitment (O'Neill, 1987). Households that are rich in land and animal resources are also short of labour, and make arrangements with neighbouring households who have plentiful labour but little land or animals (Panter-Brick, 1989). Both factors, the spatial and temporal schedule of activities and the unequal distribution of resources help 'to spread out the time of greatest labor need' (Brown, 1987).

A further point to consider is whether households are motivated to work in groups not merely to cope with time constraints but also to maximise labour efficiency. Is there benefit from an 'economy of scale', or is the work performed by labour groups 'no greater than the sum of its parts', such that 'the work done by five men in one day could be done in five days by any one of them' (Erasmus, 1965, p. 178)? Some tasks are best done by five men, as for example the work of felling trees. Some are more efficiently done on an assembly line basis, such as clearing and weeding fields, having divided related tasks amongst workers in a group. Others require the pooling of essential scarce resources, such as oxen for ploughing land (poor households with less than a pair of oxen and female-headed households without ploughmen will join forces with their neighbours). A final consideration is conviviality. Working in a group is more enjoyable, as it relieves the monotony of repetitive tasks; large parties often enliven their work with music and dancing. Cooperative arrangements also help to minimise the disputes that may arise, for instance regarding rights to divert water from communal irrigation channels to individual fields. The motivation for working in labour-groups is thus 'to do in one day a job which cannot wait five' (Erasmus, 1965), to do a job 'which exceeds the strength of a single man' (Erasmus, op. cit.), and which requires the cooperation and pooled resources of different households.

A brief examination of the relative merits and disadvantages associated with different types of labour organisation will underline the rationale governing a choice of work patterns and its relationship with a change of seasons.

### Alternative labour strategies

A finite number of labour strategies have been documented in places as diverse as the Andes (Erasmus, 1955; Guillet, 1980; Brown, 1987), the Amazon (Berte, 1988; Chibnik & de Jong, 1989), the Himalayas (Macfarlane, 1976; Messerschmidt, 1980; Panter-Brick, 1989), Africa (Swindell, 1985), Portugal (O'Neill, 1987) and the Pyrenees (Ott, 1981); see also Moore (1975) and Barlett (1980). The purpose of this chapter is to highlight their characteristics (Table 16.1) and the decision-making process for

Table 16.1. *Characteristics of different types of labour*

| Labour type | Advantages | Disadvantages |
|---|---|---|
| Household labour | Low cost<br>High quality work<br>Flexible schedule | Long time to complete tasks |
| Free assistance | Emergency help | Limited duration |
| Festive labour | Low direct cost | Indirect time and food costs<br>Low quality work |
| Reciprocal exchange | Efficiency | Reciprocal obligations |
| Hired labour | Convenience | High financial cost |

selecting between different alternatives (Fig. 16.1), namely household work, free help, festive work parties, reciprocal exchange and wage labour.

A reasonable starting point is to declare that 'households prefer to do as many agricultural tasks as possible using family labor' (Chibnick & de Jong, 1989). This obviates the costs of recruiting and feeding outside workers and ensures a high quality performance of labour. However, most households are too small to provide all the manpower required at times of a peak demand for labour.

Free help can be offered by one's neighbours, but only for a limited period of time. Households are bound by ties of mutual assistance, especially when family members experience a debilitating illness, but assistance eventually breaks down if the strain persists over time, as in cases of chronic illnesses (Thomas *et al.*, 1988; Leatherman *et al.*, 1988).

Formal and informal arrangements are made to work in labour groups when tasks must be accomplished quickly and efficiently. Festive labour is organised where a task has annual recurrence but is of limited duration. Examples are cleaning irrigation ditches in highland Peru, or building roofs for houses in Nepal. The host family avoids the expenditure of cash, tempting its neighbours to work by providing plentiful and special food and drink. Festive labour has two major disadvantages. First, an outlay of time and resources is required for the preparation of food and drink. In rural Nepal, for instance, brewing beer takes at least one day, and cigarettes must be purchased form outside towns for distribution. Second, the quality of the work suffers, because workers participate for the sake of good food and beer, and expect to get drunk. In Chile, 'the expression "This looks like mingaco [festive] work" is used to indicate disapproval of a job poorly done' (Erasmus, 1965, p. 185).

Reciprocal labour is a system of exchange in which the number of person–days labour is carefully monitored. Reciprocal exchange is favoured where farmers must complete repetitive tasks (e.g. weeding) occurring at predictable intervals, and where farmers need careful work but

224    C. Panter-Brick

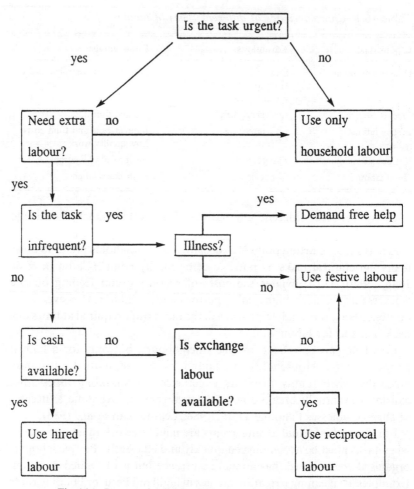

Fig. 16.1. Process of decision-making between different labour strategies (modified from Chibnick & de Jong, 1989).

do not have enough money to hire day-labourers. Reciprocal labour exchange, however, will bind the household to reciprocal obligations. In highland Nepal, where communities rely on a mixed agro-pastoral economy to achieve self-sufficiency, household size and composition can limit one's participation in all but the smallest of labour groups: receiving a party of five workers on one's land commits one to working five days on someone else's land, which is only possible if another family member assumes all other subsistence and household tasks. Households whose members cut and carried their own firewood could not spare the manpower to participate in large wood-carrying labour-groups. This is one reason why high fertility characterises the strategy of many of the rural poor who aim for sufficient manpower to make their household unit viable (Moore, 1975).

With wage labour, the principal advantage is convenience, while the main disadvantage is cost: 'Employers of hired labor spend little or no time preparing food and drink for their workers and incur no reciprocal labor obligations to their employees' (Chibnick & de Jong, 1989, p. 85), but credit must be available in cash or kind. While Erasmus (1955) and Moore (1975) thought that cash transactions would rapidly supplant cooperative labour throughout Latin America, because of their greater convenience, Chibnick & de Jong (1989) stress that reciprocal exchanges have persisted in many peasant communities, because of poverty.

The flowchart in Fig. 16.1 highlights the important factors influencing a household's process of decision-making. It also helps to predict which type of work patterns may prevail in a seasonal environment by specifying the nature of the task and the socio-ecological constraints: tasks that are urgent; tasks that occur regularly or predictably; tasks that have limited or prolonged duration; tasks for which there are scarce or available resources of cash and manpower. While the shortage of cash may handicap a household for much of the year, the shortages of time and labour are constraints that surge and wane with the seasons.

### Seasonal shifts in work patterns

Let us turn to a case study of subsistence activities in a Himalayan village and present time-allocation data to determine the extent to which farmers shift their labour strategies with the change of seasons. One can expect to find a greater reliance on reciprocal labour exchange and/or cash transactions at times when the nature of subsistence tasks is both urgent and predictable.

Fieldwork was undertaken in the village of Salme, a remote community in the foothills of the Himalayas (1870 m, Nuwakot district, north-west Nepal). The environment is highly seasonal, with a total rainfall of 4000 mm (158 inches) per year (Fig. 16.2), three-quarters of which is concentrated in the monsoon season (late June to early September). Altitude further patterns the use of time and space, with land terraced for cultivation in three successive ecological zones (low-lying irrigated paddy fields, mid-altitude maize and millet rain-fed fields, and high-altitude maize and barley fields).

A time-allocation study of work patterns was conducted throughout the period of one year on a representative sample of 87 villagers, observed for several days in the winter, spring and monsoon seasons for a total 7678

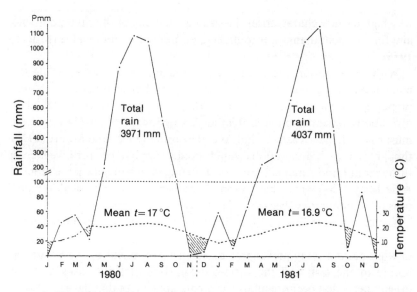

Fig. 16.2. Seasonal rainfall in Salme village (from Smadga, 1989). Shaded area indicates dry periods (rainfall less than mean temperature on a scale of 1 mm = 2 °C).

hours of minute-by-minute observation on 638 focal person-days. Two subjects were followed individually each day, by a team of four well-trained village assistants, and activities recorded continuously from 7.00 am to 6.00 pm in the winter and 6.00 am to 7.00 pm in the monsoon (11–13 hours per day). Only one caste group, the Tamang agro-pastoralists, is considered here. Data are presented for both men and women, but the sexual division of labour will not be considered in any detail.

### Seasonality of work patterns

There is a sharp increase of agricultural work input in the monsoon, from less than three hours in the winter and spring to about five hours in the monsoon (the equivalent of less than 24% to around 40% of daylight hours; $p < 0.0001$; Table 16.2 and Fig. 16.3). The day-to-day pattern of work also differs by season (see the coefficients of variation in Table 16.2). The duration of work is quite variable from day-to-day in the winter and spring, as workers combine a multitude of tasks in household maintenance, agriculture and animal husbandry. It is more uniform in the monsoon, when all household members sustain a high input of agricultural work until the planting of crops is complete. All Tamang individuals conform to this seasonal work pattern. There is no-one in the sample consistently hard-working or under-achieving.

Table 16.2. *Seasonality of agricultural work. Mean time spent working per day and percentage of total observation time*

| Season | N | Time working (hours/day) | | | Percentage of total observation time |
|---|---|---|---|---|---|
| | | Mean | (SD) | CV (%) | |
| Tamang men | | | | | |
| Early winter | 14 | 2.05 | (2.10) | 125 | 19 |
| Late winter | 21 | 0.78 | (1.13) | 177 | 7 |
| Spring | 20 | 1.26 | (1.76) | 181 | 10 |
| Monsoon | 20 | 4.70 | (2.63) | 66 | 37 |
| Tamang women | | | | | |
| Early winter | 14 | 2.65 | (1.76) | 67 | 24 |
| Late winter | 19 | 1.35 | (1.30) | 97 | 12 |
| Spring | 19 | 2.61 | (2.32) | 61 | 22 |
| Monsoon | 19 | 5.10 | (2.80) | 55 | 40 |

Excludes minutes of rest in the fields. *N* Subjects under observation.

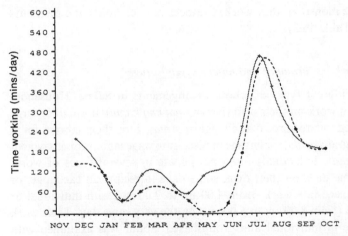

Fig. 16.3. Seasonality of agricultural work: time input for Tamang men (dotted line) and women (solid line).

Seasonality is also apparent when one takes into account the frequency with which individuals go to the fields (Table 16.3). Men and women, for instance, worked in the fields on respectively 45% and 66% of observation days in late winter compared to 85% and 87% of observation days in the monsoon (the women worked every single day in the fields from July until September 25th, when the rain subsided). The actual time-input on working days, excluding from consideration all non-agricultural days, also shows important seasonal variation. In late winter, for example, men and women averaged respectively 2.5 hours and 2.1 hours per day at work in the

Table 16.3. *Seasonality of agricultural performance on work-days – percentage of all observation days and working time (minutes) per work-day*

| Season | Days in fields (% of total) | Time working (mins/day) | |
|---|---|---|---|
| | | Mean | (SD) |
| Tamang men | | | |
| Early winter | 74 | 154 | (145) |
| Late winter | 45 | 105 | (98) |
| Spring | 50 | 152 | (163) |
| Monsoon | 85 | 331 | (153) |
| Tamang women | | | |
| Early winter | 73 | 196 | (111) |
| Later winter | 66 | 123 | (115) |
| Spring | 50 | 313 | (129) |
| Monsoon | 87 | 352 | (171) |

Excludes minutes of rest in the fields.

fields. In the monsoon, they worked respectively 5.5 hours and 5.9 hours ($p < 0.001$; Table 16.3).

## Work performance and labour organisation

There are different types of labour arrangements in Salme. Household members can work on their own (Nepali: *aaphno kaam*), in small or large groups based on reciprocal exchange (*parma*), hire their labour for a payment in grain (*jyala*), or engage in piece-rate wage labour for a payment in cash (*theka*). Individuals often participate in several types of work, working some days on their own, some days in reciprocal exchange, or combining piece-rate work (from 4.00 am to 7.00 am) with individual or wage labour (from 8.00 am to 6.00 pm) on the very same day. Their work performance in individual and group-based labour was examined with respect to the absolute duration and the consistency of time input from day-to-day, contrasting the winter and spring with the monsoon season.

The average time devoted to daily agricultural work is significantly greater when individuals work in labour groups than when working on their own ($p < 0.0001$; Table 16.4). Thus in winter and spring, men worked an average 5.3 hours per day in reciprocal exchange and only 1.8 hours per day individually, while women averaged 5.5 hours and 2.8 hours per day respectively.

The consistency of work performance is also higher in labour groups than in work for their own household. Coefficients of variation are about three times lower for days of reciprocal labour exchange than for days of

Table 16.4. *Labour organisation and work input in the fields for Tamang men and women in different seasons. Mean working time (minutes) per work-day*

| Labour context | Days working | | Time working (mins/day) | | | CV (%) | Percentage of total work |
|---|---|---|---|---|---|---|---|
| | $N$ | % | Mean | (SD) | Range | | |
| **Men: winter and spring** | | | | | | | |
| Household | 49 | 89 | 105 | (114) | 2–385 | 109 | 72 |
| Reciprocal | 4 | 7 | 316 | (93) | 245–443 | 29 | 17 |
| Hired (day) | 2 | 4 | 391 | (88) | 328–453 | 23 | 11 |
| **Monsoon** | | | | | | | |
| Household | 22 | 65 | 262 | (186) | 13–553 | 71 | 51 |
| Reciprocal | 14 | 41 | 393 | (80) | 255–452 | 20 | 49 |
| Hired (day) | 0 | 0 | — | | — | — | 0 |
| **Women: winter and spring** | | | | | | | |
| Household | 53 | 88 | 167 | (136) | 4–471 | 81 | 73 |
| Reciprocal | 8 | 13 | 331 | (109) | 197–516 | 33 | 22 |
| Hired (day) | 2 | 3 | 342 | (23) | 326–358 | 7 | 6 |
| **Monsoon** | | | | | | | |
| Household | 13 | 39 | 264 | (228) | 1–593 | 86 | 30 |
| Reciprocal | 18 | 55 | 372 | (114) | 156–565 | 31 | 58 |
| Hired (day) | 3 | 9 | 479 | (49) | 425–520 | 10 | 12 |
| Piece rate | 1 | 3 | 64 | (—) | — | — | 1 |

Excludes minutes of rest in the fields.

individual household work, and even lower for days of hired wage labour. It is apparent that participating in a labour group secures a consistently high input of labour, whereas working individually facilitates a flexible schedule of activity, alternating, for instance, agricultural tasks with animal husbandry. One strategy maximises labour efficiency, the other fosters a flexible work schedule.

Not surprisingly, household members work individually more often in the winter and spring (89% and 88% of all work-days for men and women respectively; 72% and 73% of total working time observed, men and women respectively) and more often in reciprocal and hired labour-groups in the monsoon (41% and 67% of all work-days for men and women respectively; 49% and 70% of total working time, men and women respectively) (Table 16.4).

This is quantitative evidence for a definite shift in labour organisation brought about by seasonality (Fig. 16.4). Households alternate between family-based labour and group-based labour in response to the seasonal shortages of time and manpower. Furthermore, they favour reciprocal

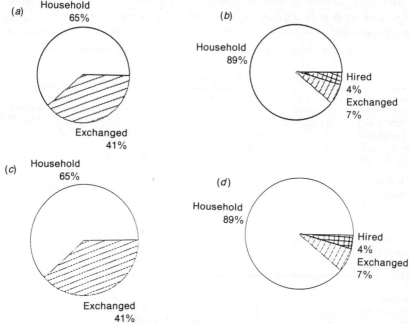

Fig. 16.4. Percentage time in household, exchanged or hired labour, by season for men and women. (*a*) men, monsoon season; (*b*) men, other seasons; (*c*) women, monsoon season; (*d*) women, other seasons.

exchanges over hiring labour. There is little wage labour in this community; payments for hired labour are mostly made in kind, and cash transactions for piece-rate work are limited to planting activities from dawn to 7.00 am in the monsoon.

## Conflict and change

The above discussion of the seasonal organisation of work patterns has focused on inter-household exchanges of time and labour, and has demonstrated a shift in labour strategies towards group-based labour and increased reciprocity in response to seasonal constraints. This supports the earlier work of Erasmus (1955) and Chibnick & de Jong (1989). Thus, households prefer to work as individual units for most of their year, as this ensures a flexible work schedule and a high-quality work performance with minimum costs and obligations. They switch to work in labour groups, however, at times of peak labour demands, since work parties constitute a more efficient way of accomplishing tasks that are both urgent and predictable in their recurrence. Peasant communities hampered by poverty, and unable to hire labour or machinery, tend to rely on reciprocal labour exchange for organising such work parties.

The logic underlying more specific strategies of labour exchange, such as the notions of equivalence between different tasks, needs to be related to the demographic, ecological and socio-economic features of a community, including patterns of residence, family size, subsistence pattern and land distribution (Moore, 1975; Panter-Brick, 1989). For example, among the Tamang of Salme, the work of adolescent girls, adult women, and pregnant or lactating mothers is deemed equal to that of men, being remunerated with the same amount of grain per day's labour. This deliberate oversight of possible differences in work performance between individuals greatly facilitates the process of inter-household labour exchange. Moreover, the work of a person per day is deemed 'equal to that of an ox', a view that simplifies the prestations of human and animal labour. These equivalences are understandable in the context of a highly endogamous, isolated community coping with multiple subsistence tasks and a shortage of manpower (Panter-Brick, 1989). They are not found in lowland villages, where men and women are paid a differential wage, and where a pool of landless workers can be hired by high-caste landowners.

Reciprocal obligations tend to weaken with the integration of peasant communities in the market economy, the local introduction of cash and the greater socio-economic differentiation between households (Painter, 1984; Moore, 1987; Leatherman *et al.*, 1988). In the words of Painter (1984, p. 289), 'a small number of people become capitalists . . . people must constantly decide between interests determined by shared ethnic identity . . . and interests determined by their relationship to capital'. In the first place, conflicts between households tend to arise with changes in the organisation of labour, as wage labour often undermines the 'moral economy' or social fabric based upon obligations of reciprocity. Fricke (1986) has a telling example of the scorn felt within a Tamang community for an entrepreneurial family who, empowered by migrant remittances, hired labour during the planting season and neglected its former network of inter-household reciprocal exchange. Second, tensions within a family unit arise from the wish of young people to retain their cash earnings for personal use. Fricke *et al.* (1990) note that some of the bitterest tensions are between parents and daughters-in-law, as the latter withhold their earnings from the common domestic fund (58% of daughter-in-laws who worked in wage labour outside their village kept most of their earnings, while only 10% of unmarried daughters did so).

Fricke *et al.* (1990) argue that 'substantial gaps impede our understanding of transition processes from family organized to wage labor market production [because] the actual processes by which these novel activities are introduced are seldom observed' (p. 284). Their detailed case study of Tamang communities in transition shows how household organisation, defined by relationships between gender and by status differences between generations, will influence an individual's participation in new forms of wage labour: the 'economic transformation is powerfully shaped by contextual factors' such as residence and proximity to markets, age or education and opportunities for employment, a woman's marital status and presence of children, which hinder a search for employment outside the village.

Wage labour and the incipient monetarisation of a local economy thus profoundly affect both strategies for household labour exchange and relationships between and within families. They also intensify another aspect of labour organisation that is heavily influenced by seasonality, namely the well-documented phenomenon of seasonal migration. 'Seasonality refers not only to . . . the timing of subsistence activities, but also to the periodicity of subsistence and wage work activities in other economic sectors' (Leatherman et al., 1988, p. 10). Seasonal migration, whereby household members leave their village temporarily to look for paid employment, illustrates well the importance of complementarity in a household's schedule of people, time and place. It is a particularly successful strategy when the period of out-migration corresponds to a slack period in subsistence activities. Absences are often prolonged, however, and in conflict with the timing of agricultural activities, resulting in great stress placed upon the management of the farm, particularly in the case of female-headed households (Leatherman et al., 1988). Tensions between gender and generations are also heightened when decisions are taken to leave the village or to retain wages earned from seasonal employment (Fricke et al., 1990).

Thus, household strategies to cope with environmental seasonality must be examined in terms of their vulnerability to short- and long-term changes in the socio-economic context. Leatherman et al. (1988) emphasise that as peasant communities become increasingly dependent on urban markets, 'seasonality surfaces in a general picture of uncertainty'. These authors focus attention on the challenges to existing seasonal strategies in the Andes, which were historically based on 'resources diversification, reciprocity and redistribution', and which 'like seasonal strategies everywhere, . . . generally seek to increase control and predictability over seasonal uncertainty and maintain flexibility in future response' (op. cit., p. 18). Important sources of conflict include the stresses associated with chronic illnesses, persistent out-migration of men for employment, and the erosion of reciprocal social networks in favour of wage labour. Some households become increasingly marginalised and vulnerable to seasonality, with net consequences for the health and nutritional status of their family members; for instance, poorer Andean families must cope with pronounced seasonal differences in energy intake and food availability (Leonard & Thomas, 1989).

An important impetus to future research will come from studies that can witness communities undergoing actual socio-economic transition and thereby document the impact of observed changes on individual behaviour and household organisation. For instance, how do existing household strategies in markedly seasonal environments cope with rapid socio-economic change (Leatherman *et al.*, 1988)? How does the switch in competing strategies based on alternatives of reciprocal exchange and wage labour relate to land distribution, family size and composition, household networks and alliances built through marriage and labour exchange (Painter, 1984; Moore, 1987; Fricke *et al.*, 1990)? And how do changes in socio-economic circumstances and family relationships affect women's activities and child health, given that emphasis on wage labour can thwart the equivalence of tasks undertaken by men, women and child-rearing mothers (by introducing differential pay-scales) or undermine the compatibility between economic and childcare activities (by frustrating arrangements for reciprocal exchange, which allow women to bring young children along to the work-place; Panter-Brick, 1989)?

### Acknowledgements

The field research in Nepal was undertaken in collaboration with the French National Centre of Scientific Research and was financed by a Leverhulme Abroad Studentship and by the Royal Anthropological Institute.

### References

Alderman, H. & Sahn, D.E. (1989). Understanding the seasonality of employment, wages, and income. In *Seasonal Variability in Third World Agriculture: The Consequences for Food Security*, ed. D.E. Sahn, pp. 81–106. Baltimore and London: The Johns Hopkins University Press.

Barlett, P. (1980). Adaptive strategies in peasant agricultural production. *Annual Review of Anthropology*, **9**, 545–73.

Berte, N.A. (1988). K'ekchi' horticultural labor exchange: productive and reproductive implications. In *Human Reproductive Behavior: A Darwinian Perspective*, ed. L. Betzig, M. Borgerhoff Mulder & P. Turke, pp. 83–95. Cambridge: Cambridge University Press.

Brown, P.F. (1987). Population growth and the disappearance of reciprocal labor in a highland Peruvian community. *Research in Economic Anthropology*, **8**, 225–45.

Chibnik, M. & de Jong, W. (1989). Agricultural labor organization in Ribereno communities of the Peruvian Andes. *Ethnology*, **28**(1), 75–95.

Erasmus, C.J. (1955). Culture, structure and process: the occurrence and disappearance of reciprocal farm labor. *Southwestern Journal of Anthropology*, **12**, 444–69.

Erasmus, C.J. (1965). The Occurrence and Disappearance of Reciprocal Farm Labor in Latin America. In *Contemporary Cultures and Societies of Latin America: A Reader in the Social Anthropology of Middle and South America and the Caribbean*, ed. D.B. Heath & R.N. Adams, pp. 173–99. New York: Random House.

Fricke, T.E. (1986). *Himalayan Households: Tamang Demography and Domestic Processes. Studies in Cultural Anthropology no. 11.* MI: Ann Arbor.

Fricke, T.E., Thornton, A. & Dahal, D.R. (1990). Family organization and the wage labor transition in a Tamang community of Nepal. *Human Ecology*, **18**, 283–313.

Guillet, D. (1980). Reciprical labor and peripheral capitalism in the Central Andes. *Ethnology*, **XIX**(2), 151–67.

Leatherman, T.L., Thomas, R.B. & Luerssen, S. (1988). Challenges to seasonal strategies of rural producers: uncertainty and conflict in the adaptive process. In *Coping with Seasonal Constraints*, ed. R. Huss-Ashmore, J.J. Curry & R.K. Hitchcock, pp. 9–20. MASCA Research Papers in Science and Archaeology 5. Philadelphia: Museum of Applied Science Centre for Archaeology, University Museum, University of Pennsylvania.

Leonard, W.R. & Thomas, R.B. (1989). Biosocial responses to seasonal food stress in highland Peru. *Human Biology*, **61**(1), 65–85.

Macfarlane, A.D. J. (1976). *Resources and Population: A Study of the Gurungs of Nepal.* Cambridge: Cambridge University Press.

Messerschmidt, D.A. (1980). Nogar and other traditional forms of cooperation in Nepal: Significance for Development. *Human Organization*, **40**(1), 40–7.

Moore, M.P. (1987). Co-operative labour in peasant agriculture. *Journal of Peasant Studies*, **2**(3), 570–91.

O'Neill, B.J. (1987). *Social Inequality in a Portuguese Hamlet: Land, Late Marriage, and Bastardy, 1870–1978.* Cambridge: Cambridge University Press.

Ott, S.J. (1981). *The Circle of Mountains: A Basque Shepherding Community.* Oxford: Oxford University Press.

Painter, M. (1984). Changing relations of production and rural underdevelopment. *Journal of Anthropological Research*, **4**, 271–92.

Panter-Brick, C. (1989). Motherhood and subsistence work: the Tamang of rural Nepal. *Human Ecology*, **17**(2), 205–28.

Swindell, K. (1985). *Farm Labour: African Society Today.* Cambridge: Cambridge University Press.

Thomas, R.B., Leatherman, T.L., Carey, J.W. & Haas, J.D. (1988). Biosocial consequences of illness among small scale farmers: a research design. In *Capacity for Work in the Tropics*, ed. K.J. Collins & D.F. Roberts, pp. 249–76. Cambridge: Cambridge University Press.

# Index

235